Archimedes

New Studies in the History and Philosophy of Science and Technology

Volume 41

Series Editor
Jed Z. Buchwald
Dreyfuss Professor of History
California Institute of Technology
Pasadena
California
USA

Archimedes, has three fundamental goals: to further the integration of the histories of science and technology with one another; to investigate the technical, social and practical histories of specific developments in science and technology; and finally, where possible and desirable, to bring the histories of science and technology into closer contact with the philosophy of science. To these ends, each volume will have its own theme and title and will be planned by one or more members of the Advisory Board in consultation with the editor. Although the volumes have specific themes, the series itself will not be limited to one or even to a few particular areas. Its subjects include any of the sciences, ranging from biology through physics, all aspects of technology, broadly construed, as well as historically-engaged philosophy of science or technology. Taken as a whole, *Archimedes* will be of interest to historians, philosophers, and scientists, as well as to those in business and industry who seek to understand how science and industry have come to be so strongly linked.

More information about this series at http://www.springer.com/series/5644

Norma B. Goethe • Philip Beeley
David Rabouin
Editors

G. W. Leibniz, Interrelations Between Mathematics and Philosophy

Editors
Norma B. Goethe
School of Philosophy
National University of Cordoba
Córdoba
Argentina

David Rabouin
CNRS
University Paris Diderot
Paris Cedex 13
France

Philip Beeley
Faculty of History
University of Oxford
Oxford
United Kingdom

ISSN 1385-0180   ISSN 2215-0064 (electronic)
Archimedes
ISBN 978-94-017-7869-5   ISBN 978-94-017-9664-4 (eBook)
DOI 10.1007/978-94-017-9664-4

Preface

The papers in this collection focus on the study of Leibniz's mathematical and philosophical thought and the interrelations between the two. They take advantage of the fact that we are today in the privileged position of being able to take a fresh look at material which has long been available in conjunction with those letters and papers recently published thanks to the remarkable efforts of the editors of the Academy Edition. With the benefit of a considerably extended textual basis, compared even to twenty years ago, we seek to examine Leibniz's mathematical practice with philosophical eyes exploring his goals and the underlying values and ideas that guided so many of his investigations.

The present volume traces its origin to a memorable workshop on the interrelationships between mathematics and philosophy in G. W. Leibniz which was organized by Mic Detlefsen and David Rabouin, and which took place at Université Paris Diderot (Laboratoire SPHERE, CNRS, UMR 7219) and at the École Normale Supérieure in Paris, 8–10 March 2010. The workshop was conceived within the framework of the "Ideals of proof" project under the direction of Detlefsen and funded by the Agence Nationale de la Recherche. Besides providing the ideal setting for discussion, that event revealed a common sentiment amongst all participants that a more in-depth study of the interrelations between these two fundamental aspects in Leibniz's thought was not only highly desirable, but also most timely on account of growing interest in the philosophy and history of mathematical practice.

Initial plans for this volume were drawn up immediately after the workshop by Norma Goethe during long hours of lively discussions over coffee and with three other participants, Richard Arthur, Philip Beeley, and David Rabouin, in the wonderful old Café Gay Lussac at the corner of Rue d'Ulm and Rue Claude Bernard. In fact, Norma Goethe, a fellow at the Lichtenberg-Kolleg (University of Göttingen) for the academic year 2009–2010, came from Göttingen with the undisclosed aim of persuading Philip and David of the timeliness of the project and invited them to join the editorial team. She should like to thank the Lichtenberg-Kolleg (University of Göttingen) and the German Research Foundation (DFG) for providing support for her participation in the workshop and the ongoing work towards the present volume which took her to Oxford and Nancy for exchanges with Philip and David. All of the editors should like to express their sincere gratitude to Jed Buchwald, editor of the

Archimedes series, for his interest in the project and also to Lucy Fleet and Mireille van Kan for their patience in the face of considerable delays in submission.

Some of the essays commissioned for this volume have grown out of papers presented in Paris, while others have been conceived and written since that time specifically for publication in this volume. All contributions have in no small measure benefitted from those three days of intense intellectual exchange and debate first in the Rue d'Ulm and then on the banks of the Seine in the Rue Thomas Mann.

The editors should like to thank all the participants of the workshop for the insights on Leibniz's mathematics which they shared and for the fruitful exchanges that were thereby made possible. Their thanks go especially to Mic Detlefsen who understood the significance of organizing such a scholarly gathering at that time and for the intellectually stimulating way in which he conducted the workshop. Particularly remembered is how his enthusiasm engendered lively interaction between all participants and how discussion continued through coffee breaks and well into the evenings.

In addition to thanking the authors who contributed to this volume, the editors should also like to thank all of the invited referees for the way in which they brought to bear their dedication to high scholarly standards. Besides those listed, we should also like to thank Marco Panza for the sound academic advice he gave. Special thanks go to Siegmund Probst for his unlimited generosity in providing all kinds of assistance to our book project. Finally, we should like to express our gratitude to Kirsti Andersen and Henk Bos for insightful exchanges and comments, wonderful conversations, and a most enjoyable time spent on the Rive Gauche after the conference was over.

Norma B. Goethe
Philip Beeley
David Rabouin

Contents

Comparability of Infinities and Infinite Multitude
Samuel Levey

Douglas M. Jesseph

Contributors

Richard T. W. Arthur Department of Philosophy, McMaster University, Hamilton, ON, Canada

Philip Beeley Faculty of History, University of Oxford, Oxford, UK

Norma B. Goethe School of Philosophy, National University of Cordoba, Cordoba, Argentina

Emily R. Grosholz Department of Philosophy, Pennsylvania State University, University Park, PA, USA

Laboratoire SPHERE, UMR 7219, CNRS—Université Paris Diderot, Paris, France

Douglas M. Jesseph Department of Philosophy, South Florida University, Tampa, FL, USA

Eberhard Knobloch Technische Universität Berlin, Berlin, Germany

Samuel Levey Philosophy Department, Dartmouth College, Hanover, NH, USA

Siegmund Probst Leibniz Archives, Hanover, Germany

David Rabouin Laboratoire SPHERE, UMR 7219, CNRS—Université Paris Diderot, Paris, France

Invited Referees

O. Bradley Bassler University of Georgia, USA

Daniel Garber Princeton University, USA

Emily R. Grosholz Pennsylvania State University & REHSEIS / SPHERE,
University Paris Diderot, Paris, France

Niccolò Guicciardini University of Bergamo, Italy

Samuel Levey Dartmouth College, USA

Jeanne Peiffer (CNRS), Centre Alexandre Koyré, Paris, France

Siegmund Probst Leibniz Archives, Hanover, Germany

Staffan Rodhe University of Uppsala, Sweden

Donald Rutherford University of California at San Diego, USA

Part I
Mathematics and Philosophy

The Interrelations Between Mathematics and Philosophy in Leibniz's Thought

Norma B. Goethe, Philip Beeley and David Rabouin

*Les Mathematiciens ont autant besoin d'estre philosophes,
que les philosophes d'estre Mathematiciens.*

Leibniz to Malebranche, 13/23 March 1699 (A II, 3, 539)

1   A Mixture of Philosophical and Mathematical Reflections and Deliberations

The aim of this collection is to explore the ways in which mathematics and philosophy (metaphysics and broader philosophical questions) are interrelated in the letters and papers of Gottfried Wilhelm Leibniz. Taking up one of his most notable expressions, the essays collected in this volume are all in some way concerned with "a curious mixture of philosophical and mathematical thought" which characterizes Leibniz's reflections and deliberations.[1] One of our principal aims in editing the present volume is to address the interrelations between mathematics and philosophy as far as possible without drawing on grand reconstructions which in the past all too often were based on insufficient evidence or what scholars conceived of as ad hoc programmatic stances, a typical example being Leibniz's so easily misunderstood pronouncement: "My metaphysics is all mathematics, so to speak, or could

---

[1] In an exchange with Basnage de Bauval, Leibniz revealed his intention to publish his correspondance with Arnauld and advanced what was to be expected from the content of his letters in these terms: "Il y aura un melange curieux de pensées philosophiques et Mathematiques qui auront peut-estre quelque fois la grace de la nouveauté"; Leibniz to Basnage de Bauval, 3/13 January 1696 (A II, 3, 121).

N. B. Goethe (✉)
School of Philosophy, National University of Cordoba, Cordoba, Argentina
e-mail: ngoethe@ffyh.unc.edu.ar

P. Beeley
Faculty of History, University of Oxford, Oxford, UK
e-mail: philip.beeley@history.ox.ac.uk

D. Rabouin
Laboratoire SPHERE, UMR 7219, CNRS—Université Paris Diderot, Paris, France
e-mail: davidrabouin2@gmail.com

© Springer Netherlands 2015
N. B. Goethe et al. (eds.), *G.W. Leibniz, Interrelations between Mathematics and Philosophy,* Archimedes 41, DOI 10.1007/978-94-017-9664-4_1

become so".[2] The difficulties presented by such reconstructions were already apparent when they emerged at the beginning of the last Century, during the second "Leibniz Renaissance" (the first having occurred in the eighteenth Century). Commentators such as Léon Brunschvicg followed an approach already adopted by the neo-Kantian philosopher Cassirer. Searching to defend him from attacks by Russell and Couturat[3], Brunschvicg criticized those who tended to confuse Leibniz's merely programmatic pronouncements with the position or rather positions which he actually maintained, which Brunschvicg termed his "real logic":

> We do not have the right to claim that Leibniz's philosophy is, properly stated, unambiguously and without ulterior motive, a *panlogism*. It would necessitate, in effect, that the relation of the predicate to the subject be achieved. In fact, the principles of 'the real logic, or a certain general analysis independent of algebra', as Leibniz put it in a letter to Malebranche, bring us back from traditional logic to differential calculus. The alternative expressed here was not completely satisfying for Leibniz in respect of his philosophical ambitions: for him, just as in the case of geometry for Descartes, differential calculus was only the most convincing 'sample' of his method, and he never gave up the project of a system of universal logic, in which the new mathematics would enter as a particular case. This is beyond doubt, but it only concerns, once more, the dream of what leibnizianism should be according to Leibniz—a dream condemned to be lost in the clouds of a tireless imagination and that for two centuries were believed to be without fruit.[4]

But despite such criticism, Brunschvicg himself (as a reflection of his time) offered his own reconstruction. He was convinced that it was possible to start from a coherent set of theses thus setting the ground for what he conceived of as Leibniz's "mathematical philosophy", while accepting that tensions and even inconsistencies might possibly remain. As a matter of fact, the use of such reading strategies was not uncommon until fairly recently amongst scholars seeking to elucidate from a variety of intellectual perspectives the way in which mathematics and philosophy are interrelated in Leibniz's thought.[5] To a certain extent, the assumptions underlying

---

[2] Leibniz to L'Hospital, 27 December 1694 (A III, 6, 253): "Ma metaphysique est toute Mathematique pour dire ainsi, ou la pourroit devenir".

[3] See Russell (1903).

[4] Brunschvicg (1912, 204). Unless otherwise stated, all the translations are ours.

[5] Concerning Leibniz scholarship in the twentieth Century, see Albert Heinekamp (1989) who distinguished three main lines of study: first, the view that focuses on the ideal of system ("à la recherche du vrai système leibnizien"); second, the defense of the "structuralist" reading ("les interprétations structuralistes"); third, the view that denies any systematic structure in Leibniz's philosophy ("refus du caractère systématique de la philosophie leibnizienne") which, according to Heinekamp, begins to be present only in the 80's. The first line of reading may be regarded as the most widely represented amongst scholars interested in studying Leibniz from the perspective of the interrelations between mathematics and philosophy. Amongst French scholars, Serres (1968) and Belaval (1960) may be mentioned as cases where the indirect impact of mid-twentieth Century foundational philosophy of mathematics and logic can be detected. One could also mention the work of G.-G. Granger (1981), who emphasizes the epistemic value of Leibniz's guiding ideas at the basis of his mathematical contributions (vis-à-vis the work of other great seventeenth Century contributions to mathematical analysis) but also sees Leibniz' mathematical work as a possible anticipation of modern non-standard analysis. For a contextual study of the development of formal logic in the late nineteenth and early twentieth Century and the exact role played by Leibniz's

such reconstructions often prevented the study of the interrelations between mathematics and philosophy in their own right.

A further difficulty with such approaches to the study of Leibniz's thought was that it motivated scholars to make sometimes arbitrary choices in his mathematical and philosophical writings without any consideration of the time and material context of production. This tendency comes to light paradigmatically in the selection of unpublished material practiced by past editors. As a matter of fact, it was precisely there where the problem started. As Couturat already noted, previous editors selected from the Leibniz's *Nachlass* the most relevant pieces to be published according to their specific intellectual interest; but unavoidably, similar objections could be made against the editor of *Opuscules et fragment inédits*.[6] While B. Russell's attempt at systematic reconstruction flatly ignored Leibniz's mathematical contributions, it is noteworthy that Cassirer and Brunschvicg, as reflected in the passage quoted above, mainly focused on the elaboration of the differential calculus taking it to be essential to understanding the interrelations between mathematics, physics, and metaphysics.[7] On the other hand, Couturat was originally motivated by G. Peano's references to Leibniz "logical insights" and anticipations to search amongst his unpublished notes for Leibniz's many experiments with "formal calculi" and other programmatic sketches related to his goal to design new working tools—which Leibniz called "characteristics"—as well as any material deemed relevant to the vision of a universal grammar, and universal mathematics with logic as the sustaining link.[8]

As noted, such lines of research by proceeding selectively led not only to the introduction of arbitrary divisions in Leibniz's writings, often ignoring chronological order, but sometimes even entailed opposing readings of one and the same section of his works. For instance, the very same texts on *analysis situs* could be interpreted either along the lines of conceptual analysis (by commentators such as Cassirer) or along the lines of formal calculus and logical theory of relations (by commentators such as Couturat).

A last difficulty presented by this time-honored approach was its pretention to propose a picture of Leibniz's philosophy as a whole. As Dietrich Mahnke emphasized already in the early 1920s, it left readers with the unfortunate impression of facing a choice between different 'paintings' of Leibniz, depending on whether or not mathematics was involved in the drawn portrait. Typical examples were, on the one hand, the project to which Mahnke gave the name "universal mathematics",

---

work as a possible anticipation of modern approaches in logic and mathematics, see Peckhaus (1997). Despite revealing historical studies, the 'logicist' trend is still represented explicitly in recent times, for instance, by Sasaki (2004, 405), who goes so far as to speak of "Leibniz's 'logicist-formalist' philosophy of mathematics".

[6] Couturat (1903), Preface.

[7] See Russell's Preface to the second edition of his book on Leibniz, Russell (1937): in composing his original book, Russell conceded that he ignored all material relevant to Leibniz's mathematical studies and contributions, but still insisted that his "interpretation of Leibniz's philosophy is still the same" as in 1900.

[8] See Couturat (1903), Preface, and Peckhaus (1997).

as dealt with in various forms by Couturat, Cassirer, and Brunschvicg and, on the other hand, the so-called "metaphysics of individuation" which he identified with commentators such as Kabitz, Sickel, and Baruzi.[9] Interestingly enough, Couturat[10] himself warned against philosophers as well as mathematicians who ignored Leibniz's recommendations that "mathematicians have just as much need to be philosophers as philosophers to be mathematicians".[11]

Once again, the elements at the basis of all these interpretations are to be found in Leibniz's writings, as well as in his rich and extensive correspondence.

The present collection of essays aims to elucidate how these different aspects in Leibniz's thought relate to each other, evolving over time as his thinking unfolds. With this aim in mind, the papers in this volume take advantage of two fortunate circumstances. First, we are today in the privileged position of being able to take a fresh look at material which has long been available in conjunction with those letters and papers most recently published by the Academy edition. With the benefit of a considerable extended textual basis we propose to look at Leibniz's mathematical practice while at the same time exploring his goals and the underlying values and ideas that guided his problem-solving activities. For example, we examine his notes and interactions with others in the process of studying mathematics in Paris under the guidance of Huygens, but we are also interested in exploring how his mathematical experience evolved, transforming his earlier philosophical views. For Leibniz, thinking unfolds and takes place in time, a fact which is beautifully reflected in his writings. The second fortunate circumstance that motivates scholarly research on the interrelations between mathematics and philosophy in Leibniz's thought relates to today's growing interest in broadening the perspective of philosophy of mathematics, so that it engages historical case-studies. The new focus on the history of mathematical practice emphasizes precisely how such practice is intertwined with philosophical ideas. The notion of a specific area of study called "philosophy of mathematics" began to develop only in the early twentieth Century as an enterprise whose main concern was to deal with growing worries about foundational issues in mathematics. This logicist project left no room for historical case studies and the institutional contextualization of mathematical practice. Instead, it focused on deductive rigor, the elaboration of predicate logic, and the axiomatic method. Leaving behind such stringent formal concerns, the field has been opening up to include the

---

[9] However, even Mahnke tried to rescue the idea of system by proposing a view which was conceived as a synthesis of both leading interpretations at his time in his book *Leibnizens Synthese von Universalmathematik und Individualmetaphysik* (Mahnke 1925).

[10] See Couturat (1901, vii): "Les philosophes, séduits à bon droit par sa métaphysique, n'ont accordé que peu d'attention à ses doctrines purement logiques, et n'ont guère étudié son projet d'une Caractéristique universelle, sans doute à cause de la forme mathématique qu'il revêtait. D'autre part, les mathématiciens ont surtout vu dans Leibniz l'inventeur du Calcul différentiel et intégral, et ne se sont pas occupés de ses théories générales sur la valeur et la portée de la méthode mathématique, ni de ses essais d'application de l'Algèbre à la Logique, qu'ils considéraient dédaigneusement comme de la métaphysique. Il en est résulté que ni les uns ni les autres n'ont pleinement compris les principes du système, et n'ont pu remonter jusqu'à la source d'où découlent à la fois le Calcul infinitésimal et la Monadologie".

[11] Leibniz to Nicolas Malebranche, 13/23 March 1699 (A II, 3, 539): "Les Mathematiciens ont autant besoin d'estre philosophes, que les philosophes d'estre Mathematiciens".

study of the work of the research mathematician, and how that work interacts with philosophical ideas and other cultural ingredients in broader historical context. This is the most welcome setting to return to the study of Leibniz, the research mathematician, who insisted upon the need to think philosophically while immersed in mathematical practice.

2 Encountering Mathematics in Paris

Although Leibniz had good political reasons for travelling to Paris in March 1672, it was the intellectual culture and above all the presence of some of the then greatest mathematical minds in Europe which persuaded him to prolong his stay, interrupted by a short visit to London, until October 1676.[12] In a letter written some two years after he had returned to Germany in order to take up his position as court counsellor and librarian in Hanover, he talks of devoting himself with an "almost limitless passion" to mathematics during those four heady years in the French capital.[13]

Leibniz's initiation to mathematics is of course associated primarily with Christiaan Huygens. On numerous occasions in later life he expresses his considerable intellectual debt to the Dutch savant.[14] However, it was some time after Leibniz's arrival in Paris before the two men actually met. Until late summer 1672, Leibniz was preoccupied with official tasks which his patron Johann Christian von Boineburg had assigned to him: the Egyptian plan, which Leibniz had himself devised in order to divert Louis XIV's military ambitions away from Europe, and the recovery of Boineburg's French rent and pension. Nonetheless, by September Leibniz had been introduced to Antoine Arnauld and Pierre de Carcavi, and soon thereafter there were encounters with the astronomers Giovanni Cassini and Ole Rømer.[15] This was the challenging intellectual environment he had long desired:

> Paris is a place where it is difficult to distinguish oneself: one finds the most capable men of the time in every kind of scientific endeavour and much effort and a little robustness is necessary in order to establish one's reputation.[16]

---

[12] Leibniz to Duke Johann Friedrich, autumn 1679 (A II, 1 (2006), 761); Leibniz to Fabri, beginning of 1677(A II, 1 (2006), 442); Leibniz to Conring, 24 August 1677 (A II, 1 (2006), 563).

[13] Leibniz to the Pfalzgräfin Elisabeth, November 1678 (A II, 1 (2006), 66).

[14] See for example Leibniz, *De solutionibus problematic catenarii vel funicularis in Actis Junii A. 1691. aliisque a Dn. I. B. propositis* (GM V, 255); *Historia et origo calculi differentialis* (GM V, 398); Leibniz to Huygens, first half of October 1690 (A III, 4, 598); Leibniz to Remond, 10 January 1714 (GP III, 606): "Il est vray que je n'entray dans les plus profondes [sc. mathematiques] qu'apres avois conversé avec M. Hugens à Paris".

[15] See Antognazza (2009, 140–141).

[16] Leibniz to Duke Johann Friedrich, 21 January 1675 (A I, 1, 491–492): "Paris est un lieu, ou il est difficile de se distinguer: on y trouve les plus habiles hommes du temps, en toutes sortes des sciences, et il faut beaucoup de travail, et un peu de solidité, pour y establir sa reputation". See also Leibniz to Gallois, first half of December 1677 (A III, 2, 293–294); Leibniz to Bignon, 9/19 October 1693 (A I, 10, 590) .

It was not until the autumn that Leibniz was able to meet with Huygens for the first time. For the Dutch savant, effectively entrusted by Colbert with the planning and organization of the Académie Royale des Sciences, this was not a meeting with an absolute stranger. Leibniz was already becoming known in the Republic of Letters as a man of prodigious learning, who besides possessing exceptional knowledge in law and philosophy was "mathematically very inclined, and well versed in physics, medicine, and mechanics".[17] But, more specifically, Huygens's attention had been drawn to the promising young man from Germany almost a year and a half before they actually met. The Bremen-born secretary of the Royal Society, Henry Oldenburg, eager to promote the growth of the new science in Germany, had spoken enthusiastically of Leibniz in his letters. In his most recent communication, he referred to Leibniz's two tracts on motion, the *Hypothesis physica nova* and the *Theoria motus abstracti*, both of which with his help had been reprinted in London under the auspices of the Royal Society in 1671. Oldenburg's description of Leibniz was clearly intended to serve as an introduction:

> He seems of no ordinary intelligence, but is one who has examined minutely what great men, both ancient and modern, have had to say about Nature, and finding that plenty of difficulties remain, has set to work to resolve them. I cannot tell you how far he has succeeded, but I dare affirm that his ideas deserve consideration.[18]

Knowing full well that Leibniz had first been motivated to write on the theory of motion after he had read the laws of motion published in the *Philosophical Transactions* by John Wallis, Christopher Wren, and Huygens himself, Oldenburg proceeded to quote a passage from Leibniz questioning the conformity of the laws presented by Huygens and Wren to the abstract concepts of motion.

3 The Mathematical Novice

It is important to recognize that the young man initiated in mathematics in the autumn of 1672 was, as Oldenburg emphasized, steeped in both ancient and modern philosophy, while having a sound knowledge of jurisprudence and Protestant and Catholic theology. By contrast, as far as mathematics was concerned, Leibniz brought with him little more than what he had been able to glean from introductory

---

[17] Boineburg to Conring, 22 April 1670, Gruber (1745, II, 1286–1287): "Leibnizio literae tuae maximo sunt solatio. Est iuvenis 24 annorum, Lipsiensis, Juris Doctor: imo doctus supra quam vel dici potest, vel credi, Philosophiam omnem percallet, veteris et novae felix ratiocinator. Scribendi facultate apprime armatus. Mathematicus, rei naturalis, medicinae, mechanicae omnis sciens et percupidus; assiduus et ardens".

[18] Oldenburg to Huygens, 28 March 1671, Hall and Hall (1965–1986, VII, 537–538/538–539): "Il ne semble pas un Esprit du commun, mais qui ait esplusché ce que les grands hommes, anciens et modernes ont commenté sur la Nature, et trouvant bien de difficultez qui restent, travaillé d'y satisfaire. Je ne vous scaurois pas dire comment il y ait reussi; j'oseray pourtant affirmer que ses pensees meritent d'estre considerées." See also Oldenburg to Huygens, 8 November 1670, Hall and Hall (1965–1986, VII, 239–240/241–242).

textbooks of Harsdörffer or Cardano and from the mathematical exploits of Thomas Hobbes—an author he had read avidly while he was in Mainz. Although he described the two tracts on motion of his youth on one occasion to Nicolas Malebranche as "the beginnings of his mathematical studies"[19], he would later generally dismiss them precisely because of their lack of sophistication in exact science. When he arrived in Paris, Leibniz was to all intents and purposes a mathematical novice.

> The desire to do justice to the favourable opinion which people had of me led me by good fortune to find new ways of analysis and to make discoveries in mathematics, although I had scarcely thought about this science before I came to France, for philosophy and jurisprudence had previously been the object of my studies from which I produced a number of essays.[20]

It is probable that the first meeting between Huygens and Leibniz took place in the Dutch savant's rooms in the Royal Library in Paris. During the course of their exchange, Leibniz mentioned with the remarkable boldness typical of his youth that he had discovered a method for summing infinite series. This method was the fruit of investigations into the Euclidean axiom "The whole is greater than its part", to which his attention had been drawn in Mainz, after reading the first part of Hobbes's *De corpore*.[21] In Chap. 8, Hobbes argues that *Totum esse maius parte*, like all geometrical axioms, must be demonstrable.[22] Already then during his service at the court of Johann Philipp von Schönborn, Leibniz had considered *Totum esse maius parte* to be reducible to the only two types of unproved truths which he considered admissible, namely definitions and identities. By the time he met Huygens he had not only succeeded in producing a syllogistic proof that every part of a given magnitude is smaller than the whole, but also, using the principle of identity, he had developed his main theorem that the summation of consecutive terms of a series of differences could be carried out over an infinite number of terms—assuming only that the expected total sum approaches a finite limit.

---

[19] Leibniz to Malebranche, end of January 1693 (A II, 2, 659): "Au commencement de mes etudes mathematiques je me fis une theorie du movement absolu, où supposant qu'il n'y avoit rien dans le corps que l'étendüe et l'impenetrabilité, je fis des regles du mouvement absolu que je croyois veritables, et j'esperois de les pouvoir concilier avec les phenomenes par le moyen du systeme des choses."

[20] Leibniz to Pellisson-Fontanier, 7 May 1691 (A I, 6, 195–196): "L'envie de me rendre digne de l'opinion favorable qu'on avoit de eue de moy, m'avoit fait faire quelques decouvertes dans les Mathematiques, quoyque je n'eusse gueres songé à cette science, avant que j'estois venu en France, la philosophie et la jurisprudence ayant esté auparavant l'objet de mes études dont j'avois donné quelques essais." See also Leibniz to Duke Johann Friedrich, 29 March 1679 (A I, 2, 155); Leibniz to Duke Ernst August, early 1680? (A I, 3, 32); Leibniz to Foucher, 1675 (A II, 1 (2006), 389); *De numeris characteristicis ad linguam universalem constituendam* (A VI, 4, 266).

[21] See Leibniz, *Historia et origo calculi differentialis* (GM V, 395).

[22] I, 8, § 25; Hobbes (1651, 72).

4 Early Successes in Paris

After listening to Leibniz's youthful deliberations, Huygens decided to put him to the test and asked him to determine the sum of the infinite series of reciprocal triangular numbers.[23]

$$1/1 + 1/3 + 1/6 + 1/10 + \text{etc}.$$

The result of this summation was already known to him, but he had not yet put this into print. Huygens also suggested that Leibniz consult two books which he had previously cited, but had not read: Wallis's *Arithmetica infinitorum* and the *Opus geometricum* of Grégoire de Saint-Vincent.

Developing a principle found in the *Opus Geometricum*, that the line segments representing terms of the geometrical progression must be considered to start from the same place, Leibniz recognized that the differences of consecutive terms are proportional to the original series.

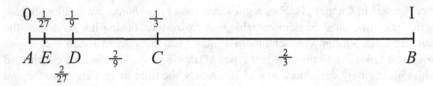

From here can be read off

$$2/3 + 2/9 + 2/27 + \ldots = 1$$

Or, more generally

$$1/t + 1/t^2 + 1/t^3 + \ldots = 1/(t-1)$$

Decisively, Leibniz was able to show how conceptually a general method could be applied. Thus, by taking AB = 1, AC = 1/2, AD = 1/3, AE = 1/4, he achieved the relation

$$1/1.2 + 1/2.3 + 1/3.4 + 1/4.5 + \ldots = 1$$

and then, multiplying by 2, produced the result which Huygens had sought, namely

$$1/1 + 1/3 + 1/6 + 1/10 + \ldots = 2$$

---

[23] See Hofmann (1974, 15).

Writing to Oldenburg on 16/26 April 1673, Leibniz does not seek to hide his joy at this early success:

> But by my method I find the sum of the whole series continued to infinity, 1/3, 1/6, 1/10, 1/15, 1/21, 1/28 etc.; indeed, I do not believe this to have been laid before the public previously for the reason that the very noble Huygens first proposed this problem to me, with respect to triangular numbers, and I solved it generally for numbers of all kinds much to the surprise of Huygens himself.[24]

Nor did Leibniz stop here, but also succeeded in obtaining the sum of the reciprocals of pyramidal numbers as well as the sum of reciprocal trigono-trigonal numbers.

$$D = 1 + 1/5 + 1/15 + 1/35 + 1/70 + \ldots = 4/3$$

The exuberance which Leibniz felt at achieving such early success—and being able to impress Huygens at the same time—can be gauged from the language he employed in what he evidently hoped would be his first mathematical publication, having already seen two letters to Oldenburg on his theory of motion published in the *Philosophical Transactions*. Most articles which appeared in the new scientific journals of the second half of the seventeenth century took the form of letters to the editor. It was therefore perfectly natural for Leibniz to set out some of his newly achieved mathematical results in a long letter to Jean Gallois, editor of the *Journal des Sçavans* and secretary of the Académie Royale des Sciences.[25] Unfortunately for Leibniz, and no doubt unbeknown to him at the time, the French journal temporarily ceased publication on 12 December 1672, that is to say, around the time his letter was sent. By the time publication was resumed on 1 January 1674, Leibniz's contribution would have been considered out of date, not least in view of the author's mathematical development during the intervening twelve months.

5 Mathematical and Philosophical Deliberations on Infinity

The *Accessio ad arithmeticam infinitorum*, as the letter to Gallois was entitled, provides evidence of the remarkable growth in Leibniz's understanding of the nature of concept of infinity compared to the views he had set out little over a year earlier in his *Theoria motus abstracti*. Whereas there he had approached the continuum ontologically, seeking to reconcile infinite divisibility with the actual existence of parts by postulating points in such a way that they could be conceived as constitutive entities, he now appeals to the argumentative force provided by genuine mathe-

---

[24] Leibniz to Oldenburg, 26 April 1673 (A III, 1, 83–89, 88): "At ego totius seriei in infinitum continuatae summam invenio methodo mea: 1/3 1/6 1/10 1/15 1/21 1/28 etc. in infinitum; quod jam publice propositum esse, vel ideo non credidi, quia a Nobilissimo Hugenio mihi primum propositum est hoc problema in numeris triangularibus; ego vero id non in triangularibus tantum, sed et pyramidalibus etc. et in universum in omnibus ejus generis numeris solvi ipso Hugenio mirante".
[25] See Bos (1978, 61).

matical proofs, such as those he had shown to Huygens, where there is an infinite progression within finite limits.

> He namely who is led by the senses will persuade himself that there cannot be a line of such shortness, that it contains not only an infinite number of points, but also an infinite number of lines (as an infinite number of actually separated parts) having a finite relation to what is given, unless demonstrations compel this.[26]

Part of what Leibniz sets out to achieve in the *Accessio* is to demonstrate that infinite number is impossible. Employing a strategy used in numerous other contemporary letters and papers, he develops his position in contrast to the position put forward by Galileo in the *Discorsi e dimostrazioni matematiche*, where infinite number, understood as the number of all numbers, is compared to unity. Galileo argued that every number into infinity had its own square, its own cube, and so on, and that therefore there must be as many squares and cubes as there are roots or integers, which however is impossible. The Pisan mathematician famously concludes from this that quantitative relations such as those of equality or "greater than" or "less than" do not apply when it comes to the infinite. That is to say, Galileo effectively negated the validity of the axiom *Totum esse maius parte* with respect to infinite numbers.

Leibniz compared Galileo's conclusion to Grégoire's negation of the validity of the axiom in horn angles in his *Opus geometricum*. In both cases, Leibniz found that it was a mistaken concept of infinity which had led to denying the universality of the axiom: "that this axiom should fail is impossible, or, to say the same thing in other words, the axiom never fails except in the case of null or nothing".[27] Precisely the universal validity of the axiom leads to the conclusion that infinite number is impossible, "it is not one, not a whole, but nothing". Employing an argument which is also found in contemporary algebraic studies, Leibniz is able to proclaim that not only is $0+0=0$, but also $0-0=0$. Consequently, an infinity which is produced from all units or which is the sum of all must in his view be regarded quite simply as nothing, about which, therefore, "nothing can be known or demonstrated, and which has no attributes".[28]

Alongside providing evidence of the relative sophistication of Leibniz's mathematical work by the end of 1672, the *Accessio* provides the earliest example of the intimate relation between philosophy and mathematics in his thought.[29] Right at the beginning, he asserts that the method of indivisibles is to be ranked among those

---

[26] Leibniz for Gallois, end of 1672 (A II, 1 (2006), 342): "Quis enim sensu duce persuaderet sibi, nullam dari posse lineam tantae brevitatis, quin in ea sint non tantum infinita puncta, sed et infinitae lineae (ac proinde partes a se invicem separatae actu infinitae) rationem habentes finitam ad datam; nisi demonstrationes cogerent."

[27] *Ibid*, 349: "at Axioma illud fallere impossibile est, seu quod idem est, Axioma illud nunquam, ac non nisi in Nullo seu Nihilo fallit, Ergo Numerus infinitus est impossibilis, non unum, non totum, sed Nihil."

[28] Leibniz, *Mathematica* (A VII, 1, 657): "Nam $0+0=0$. Et $0-0=0$. Infinitum ergo ex omnibus unitatibus conflatum, seu summa omnium esr nihil, de quo scilicet nihil potest cogitari aut demonstrari, et nulla sunt attributa." See also *De bipartitionibus numerorum eorumque geometricis interpretationibus* (A VII, 1, 227).

[29] See Beeley (2009).

things capable of vindicating the incorporeality of the mind. This assertion refers on the one hand to the geometrical method of Cavalieri , Torricelli and Roberval which had since been arithmetized by Wallis and on the other hand to one of the philosophical doctrines of his youth, namely that the immortality of the soul could be guaranteed through its location in a geometrical point. Nor was this remark at the beginning of his Paris sojourn simply a remnant of the philosophy he had brought with him from Mainz. Even in the *Système nouveau* (1695), where Leibniz considers the nature of the communication of substances and of the union between substance and body, he sees points as providing the ontological interface between the various spheres, distinguishing thereby what he calls "metaphysical points" from those of physics and mathematics.

6 Traces of a General *Ars Inveniendi*

Leibniz brings philosophical incisiveness to mathematics, analyzing concepts which contemporary mathematicians without his philosophical bent were inclined to use unreflectively. "For me the mark of imperfect knowledge," he writes to Malebranche, "is when the subject has properties of which one has not yet been able to provide a demonstration".[30] He cites the examples of the concept of a straight line employed by the geometers without having a sufficiently clear idea of what the concept involves, and of the notion of extension in respect of bodies, which clearly presupposes that there is something extended or repeated.

Conversely, Leibniz ascribed to mathematics an essential role in extending the limits of human knowledge in the context of his philosophical project of *ars inveniendi*. Shortly after he had left Paris for his new post in Hanover, he wrote that he valued mathematics solely because one could find in it "traces of a general art of invention".[31] Admittedly, Leibniz often described mathematics and indeed philosophy in terms of means to a particular end. But his evaluation of mathematics in respect to discovering new truths reflected in part the relatively recent emergence of mathematical analysis as a discipline, complementing the traditional model of a rigorously deductive science with which the concept of geometrical method had long been identified. Put simply, mathematics could now be considered to encompass both analysis and synthesis according to the ancient model of scientific method.[32]

---

[30] Leibniz to Malebranche, end of January 1693 (A II, 2, 661): "La marque d'une connoissance imparfaite chez moy, est, quand le sujet a des proprietiés, dont on ne peut encor donner la demonstration."

[31] Leibniz to the Pfalzgräfin Elisabeth?, November 1678 (A II, 1 (2006), 662): "Mais pour moy je ne cherissois les Mathematiques, que par ce que j'y trouvois les traces de l'art d'inventer en general […]." See also Leibniz to Duke Johann Friedrich, February 1679 (A II, 1 (2006), 684).

[32] See Leibniz, *De arte characteristica inventoriaque analytica* (A VI, 4, 321): "Duobus maxime modis homines inventores fieri deprehendo, per Synthesin scilicet sive Combinationem et per analysin; utrumque autem vel facultati natura usuve comparatae, vel methodo debere." See also *ibid.* (A VI, 4, 329).

Moreover, these two basic paths to new knowledge would be further enhanced and vastly extended by the implementation of a suitable, that is to say exact system of symbols which would mirror not only the structure of concepts but also thought itself, which could thereby be effectively replaced by a symbolic calculus.

The importance of such a calculus is formulated explicitly in his remarks on George Dalgarno's *Ars signorum*, probably written after his return to Paris following his first visit to London in 1673. Seeking to proceed further than contemporary exponents of artificial languages, he describes his universal character as being among "the most suitable instruments of the human mind, having namely an invincible power of invention, or retention, and judgment. Then this will accomplish in all matters of things, which arithmetical and algebraic symbols accomplish in mathematics".[33]

Building on his early fascination with the art of combinations, Leibniz recognized that a synthetic or deductive model proceeding systematically from simple elements, representing the alphabet of human thought[34], would not only serve as a suitable means of presenting existing knowledge, but also of acquiring entirely new knowledge. In this way, *ars combinatoria* could be understood properly as an important part of the art of invention. In his letter to Jean Gallois of December 1678, he writes:

> I am more and more convinced of the utility and reality of this general science, and I see that few people have grasped its scope. But in order to make this science easier and so to speak sensible, I want to employ the characteristic of which I have spoken to you a number of times, and of which algebra and arithmetic are just samples. This characteristic consists in a certain writing or language, (for whoever has the one may have the other) which corresponds perfectly to the relations of our thoughts. This science will be quite different from everything which one has planned up to now. For the most important part has been overlooked, which is that the characters of this writing must be conducive to discovery and judgment as they are in algebra and arithmetic.[35]

Evidently, one of the by-products of Leibniz's early work on mathematics, and particularly algebra, during his stay in Paris was to recognize the full potential for deriving a symbolic calculus in order to extend human knowledge. On occasion Leibniz describes his *characteristica universalis* as a "universal algebra", with whose help it would in his view be just as easy to make discoveries in ethics, physics or

---

[33] Leibniz, *Zur Ars signorum von George Dalgarno* (A VI, 3, 170): "sed vera Characteristica Realis, qualis a me concipitur, inter [ap]tissima humanae Mentis instrumenta censeri deberet, [invin] cibilem scilicet vim habitura et ad inveniendum, et ad retinendum et ad dijudicandum. Illud enim efficient in omni material, quod characteres Arithmetici et Algebraici in Mathematica." See also Antognazza (2009, 162).

[34] See Leibniz, *De alphabeto cogitationum humanarum* (A VI, 4, 271–272).

[35] Leibniz to Gallois, 19 December 1678 (A III, 2, 570): "Je suis confirmé de plus en plus de l'utilité et de la realité de cette science generale, et je voy que peu de gens en ont compris l'étendue. Mais pour la rendre plus facile et pour ainsi dire sensible; je pretends de me servir de la Characteristique, dont je vous ay parlé quelques fois, et dont l'Algebre et l'Arithmetique ne sont que des échantillons. Cette Characteristique consiste dans une certaine ecriture ou langue, (car qui a l'une peut avoir l'autre) qui rapporte parfaitement les relations des nos pensées. Ce charactere seroit tout autre que tout ce qu'on a projetté jusqu'icy. Car on a oublié le principal qui est que les caracteres de cette écriture doivvent servir à l'invention et au jugement, comme dans l'algebre et dans l'arithmetique".

mechanics as it is in geometry.[36] An essential part of this consideration is that the rigor of mathematics will also apply here, enabling us to have no less certainty about God and the mind than about figures and numbers. In this way, Leibniz suggests, inventing machines would be no more difficult than constructing a problem in geometry. He expresses the full promise of universal character in this context in his letter to Oldenburg of 28 December 1675:

> This algebra (of which we deservedly make so much) is only part of that general system. It is an outstanding part, in that we cannot err even if we wish to, and in that truth is as it were delineated for us as though with the aid of a sketching machine. But I am truly willing to recognize that whatever algebra furnishes to us of this sort is the fruit of a superior science which I am accustomed to call either Combinatory or Characteristic, a science very different from either of those which might at once come to one's mind on hearing those words [...] I cannot here describe its nature in a few words, but I am emboldened to say that nothing can be imagined which is more effective for the perfection of the human mind.[37]

But, by reading this kind of declarations, one should also keep in mind Brunschvicg's warning and not confuse "the dream of what leibnizianism should be according to Leibniz" with his "real logic". Indeed the same letter to Oldenburg begins with an important *caveat*: "we seem to think of many things (though confusedly) which nevertheless imply contradiction". Here again the motivation comes from mathematics, the basic example of contradictory notion mentioned being precisely the one presented in the *Accessio ad arithmeticam infinitorum*: "the number of all numbers" (A II, 1, 393). This a typical example of a joining together of apparently simple ideas (unit and addition), which produces an impossible object (the sum of all units or "number of all numbers"). As emphasized by the *De synthesi et analysi universali*, one must then take care that "the combinations do not become useless through the joining-together of incompatible concepts". If the *universal character*, based on the constitution of an "alphabet of human thoughts" and the full development of an *ars combinatoria*, is the goal to obtain, one should not forget that it implies nothing less than a complete analysis of human thoughts. Before reaching this goal, which may well be inaccessible to finite human beings, one has to be very cautious with symbolic manipulations, keeping in mind that they must be complemented by demonstrations of possibility : "one must be especially careful, in setting up real definitions, to establish their possibility, that is, to show that the concepts from which they are formed are compatible with each other".[38] Since the main field in which Leibniz developed such an "analysis of thoughts" and such a work on defi-

---

[36] Leibniz to Mariotte, July 1676 (A II, 1 (2006), 424): "ce seroit pour ainsi dire une algebre universelle, et il seroit aussi aisé d'inventer en morale, physique ou mechanique, qu'en Geometrie".

[37] Leibniz to Oldenburg, [18]/28 December 1675 (A III, 1, 331): "Haec algebra, quam tanti facimus merito, generalis illius artificii non nisi pars est. Id tamen praestat, ut errare ne possimus quidem, si velimus, et ut veritas quasi picta velut machinae ope in charta expressa deprehendatur. Ego vero agnosco, quicquid in hoc genere praebet algebra, non nisi superioris scientiae beneficium esse, quam nunc combinatoriam, nunc characteristicam appellare soleo, longe diversam ab illis, quae auditis his vocibus statim alicui in mentem venire possent".

[38] Leibniz, *De synthesi et analysi universali seu Arte inveniendi et judicandi* (A VI, 4, 540).

nitions was precisely mathematics, this will be enough to indicate the complexity of the interrelations between the various domains under consideration.

7   Presentation of the Collection of Essays

As should be clear from the historical sketch proposed above, mathematics and philosophy evolved *in tandem,* fruitfully interacting in Leibniz's work, influencing each other in multifarious ways throughout the different periods of his intellectual life. Yet relatively few studies have been devoted to the investigation of these complex interrelations. One reason may be the fact that Leibniz's scholarship has for too long been rather compartmentalized, with the study of metaphysics on the one side, and the study of mathematics on the other, each of these pursuits involving technicalities of its own which would require it to be placed within the context of the time. One could also invoke the changing perceptions in the history of mathematics itself, which in the last thirty years has moved away from "internalist" accounts advocated by the founders of the discipline. The availability or rather lack of availability of most of Leibniz's mathematical papers of course did not help. Until fairly recently, commentators were largely reliant on articles which Leibniz published during his lifetime or the few texts which in intervening years found their way into print. Over the last 20 years things have changed dramatically for the better. Progress in the edition of the Academy Edition of Leibniz's letters and papers has made available to readers many of the previously unpublished drafts or letters long hidden from public view. Material edited in Series VII (Mathematical Papers) as well as in Series III (Mathematical and Scientific Correspondence), not forgetting Series I (General and Political Correspondence), Series II (Philosophical Correspondence), and Series VI (Philosophical Papers) shows just how closely related Leibniz's philosophical and mathematical reflections sometimes were.

As already noted, together with the newly available material, the papers in this collection also take advantage of the growing interest amongst philosophers and historians of mathematics in addressing the work of the research mathematician, his mathematical practice in specific institutional contexts, often in exchange with others. Thus the scholar enters the workshop of the mathematician to explore underlying values, guiding ideas, methods and working tools, a strategy which in the case of the mathematician-philosopher Leibniz seems most promising. In his paper, Philip **Beeley** invites us to meet Leibniz, the philosopher mathematician who could not help but think as a mathematical philosopher. The paper shows Leibniz's great concern to account for the explanatory power of the mathematical sciences as applied to our understanding of the natural world, an interest that can be traced to earlier writings from the Mainz Period (before arriving in Paris). Leibniz's ultimate motivation was his recognition of the usefulness of the mathematical sciences with a view to the improvement of the human condition. The deep interconnection Leibniz saw between theory and practice inspired him to discuss mathematical working tools such as the notion of "negligible error" used in justifying infinitesimal techniques in connection with practical matters and its applicability in the natural world. In his discussion of "negligible error," Leibniz revisits his early interest in Archimedean ideas further developed by his later mathematical studies, a conjunction which is not divorced from its special place in the search for wisdom. Discussions such as this and related issues reveal that the dialogue between philosophy and mathematics was not just a novelty brought about by his mathematical studies in Paris (1672–1676).

The emphasis on pragmatic considerations in Leibniz's mathematical practice allows us to trace an important evolution in his thought. Careful scholarship reveals that earlier versions of this great project of an *ars combinatoria*, which if fully realized would have led to establishing an "alphabet of human thoughts", and which a very young Leibniz once assumed was objectively possible, were abandoned. In his paper "The difficulty of being simple", David **Rabouin** shows that with the start of his studies in Paris, Leibniz was motivated seriously to question the feasibility of such a project. In particular, the study of mathematics played a decisive role in this evolution. The *Accessio ad arithmeticam infinitorum* and the demonstration of the impossibility of a "number of all numbers", as already noted, as well as his work on the "arithmetic quadrature of the circle" culminating in another demonstration of impossibility, and the study in number theory, gave Leibniz new insight into crucial questions about the possibility (and impossibility) of notions. Accordingly, the form that an "analysis of human thoughts" should take evolved considerably during this period; Leibniz's mathematical practice transformed his way of engaging with mathematical concepts.

The question of why mathematics not only applies to the natural world but also helps us to find explanations of natural phenomena was also of great importance to Leibniz. He sought a middle pathway between Bacon's empiricism and the rationalism of Descartes as he framed his conception of scientific method. As Emily **Grosholz** argues in "Leibniz and the Philosophical Analysis of Time," he came to think that mathematics and experience were limited approaches to the study of nature when taken in isolation, and thus should be considered *in tandem*. Leibniz calls upon metaphysics, in particular the principles of Continuity and Sufficient Reason, to play a harmonizing role, as he sought to answer the question about how the two scientific activities (theoretical analysis and empirical compilation) should be combined in practice. She argues in particular that metaphysical principles play a substantive role in his account of time. Another remarkable aspect that comes to the fore in Grosholz's study is a conception of scientific research which involves a set of values, perhaps the most important of which is the idea that the use of mathematics applied to nature requires careful philosophical reflection.

The complete and carefully designed study of *De quadratura arithmetica circuli ellipseos et hyperbolae* (1675/1676) which Leibniz himself originally intended to publish, was meticulously edited by Eberhard Knobloch and first published in 1993. This edition offered a welcome occasion for a revival of interest in the study and assessment of Leibniz's views on infinitesimals, including Leibniz's use and interpretation of the role of "syncategorematical" expressions.[39] As the historian of

---

[39] Concerning this issue, see the material gathered in Jesseph and Goldenbaum (2008) .

mathematics Henk Bos (2001) showed in his study of the role of exactness in Descartes' work on geometry, discernible just under the surface of mathematical working tools lie implicit epistemic values that operate in mathematical practice, but often are never made explicit by the mathematician. Thus, later scholarly debates concerning the relevant values cannot be easily settled. In his essay "Analyticité, équipollence et théorie des courbes chez Leibniz", Eberhard **Knobloch** likewise approaches Leibniz's mathematical writings by studying the way in which he conceived of the relationship between "geometricity"and "analyzability". He also considers the way that Leibniz's thought evolved throughout his mathematical research. For instance, Leibniz starts out by borrowing notions from Cartesian geometry, but reworks them while progressively transforming their use and meaning. As an illuminating example of this process, Knobloch discusses the Leibnizian notion of "equipollence" which reveals itself as one of the key tools for expanding the range of objects (curves) that can be treated mathematically by using his new methods.

Epistemic values also play a key role in Leibniz's invention of the differential calculus. The philosophical project of a "general character", which turned out to be one of Leibniz's most fruitful guiding ideas, was central to the search for a symbolic calculus able to express techniques stemming from infinitesimal analysis in an economical way. This may be part of the reason why Leibniz was often unconcerned about acknowledging results previously established by other mathematicians. In his essay on "Leibniz as second inventor", Siegmund **Probst** delivers a careful investigation, based on recently edited material, of the relationship between Leibniz and his predecessors, especially Isaac Barrow and Pietro Mengoli. Although some results were the outcome of Leibniz's intensive study of the relevant sources of the time which often overlap concerning the consideration of specific topics, Probst argues, the Hanoverian philosopher-mathematician was probably more concerned with the introduction of new methods and a new kind of access to those results, which only a symbolic calculus operating at a higher level of abstraction could provide.[40] To take up Leibniz's own triumphant words: "Most of the theorems of the geometry of indivisibles which are to be found in the works of Cavalieri, Vincent, Wallis, Gregory, and Barrow are immediately evident from the calculus".[41]

The concept of infinity and its historical adjunct, the concept of continuity, constitutes in many ways an important focus of the meeting of mathematics and philosophy in Leibniz. Philosophical reflections on the infinitely small and the infinitely large abound in his letters and papers. Indeed the concept of the continuum effectively constitutes a thread through the whole of his philosophical thought from the *Hypothesis physica nova* of his youth through to the doctrine of monads of his

---

[40] For a discussion of the different "levels of abstraction" in Leibniz's mathematical practice, see Breger (2008b) and, in particular, in connection with the present idea, see Breger (2008a, 193): "[…] it was only by proving many theorems and gaining experience with the new material that Leibniz arrived at the higher level of abstraction from which he was able to recognize and explicitly formulate the rules of calculus".

[41] Leibniz, *Analyseos tetragonisticae pars tertia* (A VII, 5, 313): "Pleraque theoremata Geometriae indivisibilium quae apud Cavalerium, Vincentium, Wallisium, Gregorium, Barrovium extant statim ex calculo patent".

maturity. Although he tells us already in *De quadratura arithmetica circuli* that metaphysical considerations in respect of the infinite are of no consequences when mathematical rigor can be shown to obtain, he nonetheless recognizes that precisely the concept of *infinite parvum* cannot of itself be above philosophical analysis if it is to serve its function of explaining the applicability of the infinitesimal calculus to those natural phenomena which are its object.[42]

Since Leibniz developed and promoted infinitesimal analysis and since also he claimed to be an ardent supporter of the existence of actual infinite in nature[43], one might think that he was furthermore an ardent supporter of actual infinite entities in mathematics. But this is not what the sources tell us. Quite on the contrary, Leibniz regularly insists on the fact that he does not believe in actual infinite in mathematics. This raises many questions which have long remained hidden in past reconstructive approaches and which only now are being raised. First, what is exactly his view, or perhaps better, what were his views, on the ontological status of the infinite in mathematics, be it the infinitely large or the infinitely small? Second, how can we reconcile two apparently incompatible theses according to which Leibniz on the one hand supported the existence of an actual infinite in nature and on the other hand denied its existence in mathematics? Is it not the case that we have to accept that there is an infinite number of things in the world? And if so, how can we express this infinity?

In "Leibniz's Actual Infinite in Relation to his Analysis of Matter", Richard **Arthur** tackles precisely the last problem mentioned, namely, how to understand why Leibniz denied the existence of an infinite number in mathematics while positing actual infinity in Nature—such as in the infinite division of matter or in the plurality of simple substances. First of all, he sets out to defend Leibniz's views on the mathematical infinite as a fiction against accusations of inconsistency raised in recent literature. Such claims are often based on the anachronistic point of view of our modern "Cantorian" theory of the infinite. In the remaining part of the paper, Arthur confronts a dilemma already raised by Russell: if infinite plurality is just a fiction, depending on the way we perceive things, then there appears to be no way to assert that there is an infinite plurality of substances or that matter is actually divided into an infinity of parts. If, on the contrary, there is a real, mind-independent,

---

[42] See Leibniz to Schmidt, 3 August 1694 (A I, 10, 499): "Novum Calculi Analytici genus a me in Geometriam introductam [...] Usum inprimis habet ad ea analysi subjicienda, in quibus quantitates finitae determinantur interveniente aliqua consideratione infiniti, quemadmodum saepe praesertim cum Geometria applicatur ad naturam. Ubique enim infinitum Naturae operationibus involvitur. See also Leibniz to Kochański, 10/20 August 1694 (A I, 10, 513–514); Leibniz to the Electress Sophie for the Duchess Elisabeth Charlotte of Orléans, 28 October 1696 (A I, 13, 85): "Et c'est une chose estrange, qu'on peut calculer avec l'infini comme avec des jettons, et que cependant nos Philosophes et Mathematiciens ont si peu reconnu combien l'infini est mêlé en tout".

[43] Leibniz to Foucher, end of June 1693 (A II, 2, 713): "Je suis tellement pour l'infini actuel, que au lieu d'admettre que la nature l'abhorre, comme l'on dit vulgairement, je tiens qu'elle l'affecte partout, pour mieux marquer les perfections de son auteur. Ainsi je crois qu'il n'y a aucune partie de la matiere, qui ne soit, je ne dis pas divisible mais actuellement divisée, et par consequent la moindre particelle doit estre considerée comme un monde plein d'une infinité de creatures differentes."

infinite plurality of substances, or infinite plurality of parts of matter, then one must acknowledge infinite pluralities which are not fictions and which would correspond to the actual infinite wholes that Leibniz wants to exclude from mathematics. The solution to the dilemma, Arthur argues, is that one should not confuse the plurality itself with its perception as a unity. On this basis, it is possible to understand how the infinite plurality of parts of matter is reconcilable with the infinite plurality of substances, assuming, as Leibniz repeatedly argues, that these parts are real .

In "Comparability and Infinite Multitude in Galileo and Leibniz", Sam **Levey** revisits the contrasted positions of those two thinkers on the status of "infinite multitude". Galileo's paradox, which shows that one infinite multitude can be put in one-to-one correspondence with another even when one is a proper part of the other (such as in the case of natural numbers and their squares), was instrumental in Leibniz's reflections. In the *Accessio ad arithmeticam infinitorum*, as we already mentioned, he argues against the Pisan mathematician that infinity should not be compared with unity (which is "equal" to its powers), but with zero or "nothing". According to Leibniz, this means that there is no such thing as an infinite number, and more generally that a mathematical infinite cannot be considered as a "whole". Hence emerges a way of saving Euclid's axiom ("the whole is greater than the part") which enters as essential ingredient in Galileo's paradox. This is, however, only one amongst a number of strategies to save the axiom. Another possibility, often ascribed to Galileo himself, is that the infinite falls outside of the realm of quantifiable entities (*quanti*). Levey reexamines these interpretations in detail in order to assess the pertinence of Leibniz's strategy and its strength.

Finally, in his paper "Leibniz on The Elimination of Infinitesimal", Douglas **Jesseph** studies the status of infinitesimal quantities in Leibniz. As already noted, recent scholarly research, inspired by the rediscovery of *De quadratura arithmetica circuli* has set emphasis on revisiting the so called "syncategorematical" interpretation. According to this view, infinitesimals are "useful fictions" in the sense that they can be eliminated through a paraphrase involving only finite quantities. Following the seminal investigation published by Henk Bos in 1974, Jesseph argues that this is only one amongst two strategies to "find truth in fiction". He proposes to contrast each strategy as a "syntactic" (or "proof theoretic") and a "semantic" (or "model theoretic") approach. In the semantic approach, one seeks to show that, even if reference to infinitesimals cannot be eliminated from the mathematical discourse, it will never lead from truth to falsehood. The paper gives an example of these two strategies in Leibniz's texts and seeks to explain why they had to coexist in his mathematical practice.

References

Antognazza, Maria Rosa. 2009. *Leibniz: An intellectual biography*. Cambridge: Cambridge University Press.
Beeley, Philip. 2009. Approaching infinity: Philosophical consequences of Leibniz's mathematical investigations in Paris and thereafter. In *The philosophy of the young Leibniz,* eds. Mark Kulstad, Mogens Lærke, and David Snyder, 29–47. Stuttgart: Steiner Verlag.

Belaval, Yvon. 1960. *Leibniz critique de Descartes*. Paris: Gallimard.

Bos, Henk J. M. 1974. Differentials, higher-order differentials, and the derivative in the Leibnizian calculus. *Archive for History of the Exact Sciences* 14:1–90

Bos, Henk J. M. 1978. The influence of Huygens on the formation of Leibniz' ideas. In *Leibniz in Paris (1672–1676)*, eds. Albert Heinekamp, Dieter Mettler, and Ingeborg von Wilucki, 2 vols, I, 59–68. Wiesbaden: Steiner Verlag.

Bos, Henk J. M. 2001. *Redefining geometrical exactness: Descartes' transformation of the early modern concept of construction*. New York: Springer Verlag.

Breger, Herbert. 2008a. Calculating with compendia. In *Infinitesimal differences*, eds. Jesseph and Goldenbaum, 185–198.

Breger, Herbert. 2008b. The art of mathematical rationality. In *Leibniz, What kind of rationalist?* ed. Marcelo Dascal, vol. 13, 141–152. Dordrecht: Springer (Series: Logic, Epistemology and the Unity of Science).

Brunschvicg, Léon. 1912. *Les étapes de la philosophie mathématique*. Paris: Félix Alcan.

Couturat, Louis. 1901. *La Logique de Leibniz d'après des documents inédits*. Paris: Félix Alcan.

Couturat, Louis. 1903. *Opuscules et Fragments Inédits de Leibniz, Extraits des manuscrits de la Bibliothèque Royale de Hanovre*. Paris: Félix Alcan.

Granger, Gilles-Gaston. 1981. Philosophie et Mathématique Leibnizienne. *Revue de Métaphysique et De Morale* 86 (1): 1–37.

Gruber, Johann Daniel. 1745. *Commercii epistolici leibnitiani*. 2 vols. Hanover: I. W. Schmid.

Hall, A. Rupert, and Marie Boas Hall., eds. *Correspondence of Henry Oldenburg*. 13 vols. Madison: University of Wisconsin Press (and successors, 1965–86).

Heinekamp, Albert. 1989. L'Etat actuel de la recherche leibnizienne. *Etudes philosophiques* 2:139–160

Hobbes, Thomas. 1651. *Elementorum philosophiæ sectio prima de corpore*. London: Andrew Crook.

Hofmann, Joseph Ehrenfried. 1974. *Leibniz in Paris 1672–1676. His growth to mathematical maturity*. Cambridge: Cambridge University Press.

Jesseph, Douglas, and Ursula Goldenbaum, eds. 2008. *Infinitesimal différences. Controversies between Leibniz and his Contemporaries*. Berlin: Walter De Gruyter.

Leibniz, Gottfried Wilhelm. *Leibnizens Mathematische Schriften*, ed. C. I. Gerhardt, 7 vols. Berlin: A. Asher and Halle: H. W. Schmidt (1849–63 [quoted as **GM**]).

Leibniz, Gottfried Wilhelm. *Die Philosophischen Schriften und Briefe von Gottfried Wilhelm Leibniz*, ed. C. I. Gerhardt, 7 vols. Berlin: Weidman (1875–90 [quoted as **GP**]).

Leibniz, Gottfried Wilhelm. 1923. *Sämtliche Schriften und Briefe*, ed. Prussian Academy of Sciences (and successors); now: Berlin-Brandenburg Academy of Sciences and the Academy of Sciences in Göttingen. 8 series, Darmstadt (subsequently: Leipzig); now: Berlin: Otto Reichl (and successors); now: Akademie Verlag. [quoted as **A**].

Leibniz, Gottfried Wilhelm. 1993. *De quadratura arithmetica circuli ellipseos et hyperbolae cujus corollarium est trigonometria sine tabulis,* ed. Eberhard Knobloch. Göttingen: Vandenhoeck and Ruprecht.

Mahnke, Dietrich. 1925. *Leibnizens Synthese von Universalmathematik und Individualmetaphysik*. Halle: Max Niemeyer (*Jahrbuch für Philosophie und Phänomenologische Forschungen*), 305–612. Reprint, Stuttgart, 1964.

Peckhaus, Volker. 1997. *Logik, Mathesis universalis and allgemeine Wissenschaft. Leibniz und die Wiederentdeckung der formalen Logik im 19. Jahrhundert*. Berlin: Akademie Verlag.

Russell, Bertrand. 1903. Recent work on the philosophy of Leibniz. *Mind* 12 (46): 177–201.

Russell, Bertrand. 1937. *A critical exposition of the philosophy of Leibniz. New impression with a new preface*. 2nd ed. London: George Allen Unwin Ltd.

Sasaki, Chikara. 2004. *Descartes's Mathematical Thought*. Dordrecht: Kluwer.

Serres, Michel. 1968. *Le Système de Leibniz et ses modèles mathématiques*. Paris: P.U.F.

Leibniz, Philosopher Mathematician and Mathematical Philosopher

Philip Beeley

Of the numerous constants in Leibniz's philosophy, stretching from his intellectually formative years in Leipzig and Jena through to the mature writings of the *Monadology* conceived largely in Hanover and Berlin, few are as remarkable as his conviction that a firm understanding of the concepts of unity and infinity ultimately provide the key to developing sound metaphysics. When he famously wrote to Gilles Filleau des Billettes in December 1696 that his fundamental considerations rest on two things, namely unity and infinity[1], he simultaneously situated his metaphysics on the one side in the philosophical tradition of Aristotle and Thomas Aquinas and on the other side in the new mechanistic approach of Galileo and Descartes. For infinity and the intimately related concept of continuity, while having an ancient philosophical tradition of their own, were for him always also the key to harmonizing or at least connecting the three central pillars of his system: mathematics, physics, and indeed metaphysics itself. Even before his progression from mathematical novice, who had probably read little more on the subject than Daniel Schwenter's *Erquickstunden*, and some of Girolamo Cardano's *Practica arithmeticae generalis*[2], to becoming one of the most productive mathematical thinkers of his day, Leibniz recognized that mathematics was the foundation on which modern scientific and technological advances leading to an improvement of the human condition firmly rested[3]. Avoiding the negative theological consequences of Cartesianism and atomism, his metaphysics would ultimately conform more radically to the conceptual foundations of the mathematical sciences than the philosophical positions of any of his contemporaries.

---

[1] Leibniz to Des Billettes, 4/14 December 1696, Leibniz 1923 (cited hereafter as 'A') I, 13, 90: "Mes Meditations fondamentales roulent sur deux choses, Sçavoir sur l'unité, et sur l'infini."

[2] See Hofmann 1974, pp. 3–4.

[3] See Leibniz, De republica literaria, A VI, 4, 432, 438.

P. Beeley (✉)
Faculty of History, University of Oxford, Oxford, UK
e-mail: philip.beeley@history.ox.ac.uk

© Springer Netherlands 2015
N. B. Goethe et al. (eds.), *G.W. Leibniz, Interrelations between Mathematics and Philosophy,* Archimedes 41, DOI 10.1007/978-94-017-9664-4_2

1   Rivers and Motion

Evidence of Leibniz's recognition of this fundamental role of mathematics is to be found in his earliest extant piece of writing on the theory of motion. In *De rationibus motus*, drawn up under remarkable circumstances in 1669, he describes mathematics metaphorically as the source of the mixed sciences, on which so much useful knowledge depends:

> We therefore set out the foundations of motion, such as they are in the pure state of nature, for they are neither strengthened by demonstrations, nor embellished by logical conclusions, though these conclusions be infinite and manifest; for the sources of the arts, though they are accustomed because of a certain aridity and simplicity to displease the fastidious, flow from thence through continuous descent to the richest rivers of the sciences, and finally as if into an ocean of uses and applications.[4]

It is surely no accident that this metaphor was conceived during Leibniz's first encounter with recent scientific work from France and England on the theory of motion. In August 1669, he spent three weeks in the spa town of Bad Schwalbach in Hesse accompanying his patron Johann Christian von Boineburg, who regularly spent part of the summer taking waters away from the stresses and strains of court life in Mainz. While he was there, Leibniz was lent the recently-published April 1669 issue of the *Philosophical Transactions* by a friend of Boineburg's, the Kiel law professor Erich Mauritius. Geographically far removed from the meetings of the leading scientific community in Europe, he was able to read that the discovery of the laws of motion was a central concern to members of the Royal Society of London. The April issue contained Christiaan Huygens's contribution to the debate, which controversially had been suppressed from earlier publication, because the publisher, Henry Oldenburg, had considered Huygens's rules to be largely identical to those of the English mathematician Christopher Wren[5].

During the next two years, Leibniz would adopt quite a different approach to work on the laws of motion than that which he found while reading the *Philosophical Transactions* in the idyllic surroundings of Bad Schwalbach.[6] But for the moment it was primarily the geography of the region itself which arrested the young philosopher's mind. Bad Schwalbach lies on a tributary of the river Aar, which flows into the river Lahn. Its waters flow into those of Germany's great mercantile artery, the Rhine, and they in turn flow into the North Sea. Leibniz's surroundings suggested a metaphor for the relations between theory and practice, mathematics,

---

[4] Leibniz, *De rationibus motus* § 7, A VI, 2, 160: "Ordiamur igitur proferre Fundamenta motuum, qualia sunt in puro naturae statu, neque tamen munita demonstrationibus, neque ornata consectariis, infinitis tamen illis et illustribus; nam fontes artium, ut ariditate quadam et simplicitate delicatis displicere solent, ita decursu perpetuo in uberrima scientiarum flumina, denique quoddam, velut mare usus ac praxeos excrescunt."

[5] Christiaan Huygens's 'Regulae de motu corporum ex mutuo impulsu' were published under the English title 'A Summary Account of the Laws of Motion' in *Philosophical Transactions* Huygens 1669, pp. 927−928. (Oldenburg's account of the controversy is printed pp. 925−9277.) The laws had been published before in French in the *Journal des Sçavans*, 18 March 1669, pp. 22−24.

[6] See Garber 1995, especially pp. 273−281; Beeley 1999, pp. 134−135.

the mathematical sciences, and their applications which would hold true throughout the whole of his philosophical career. Indeed, as I will seek to show in the course of this chapter, the river metaphor only begins to acquire its full significance towards the end of his life, when so many of his theories and ideas contained in his innumerable drafts, memoranda, and letters were falling into place. Just as there is remarkable constancy in the theoretical importance of the concepts of unity and infinity in his thought, so, too, in the systematic importance he attaches to mathematics. And as I also hope to demonstrate, the concept of infinity serves as a vital link between the two.

2 Three Central Considerations: Utility, Conciliation and Rigour

It cannot be emphasized enough that among seventeenth-century and early eighteenth-century Enlightenment philosophers Leibniz stands out as having thought more than any other about the need to provide a rational account of the successful application of mathematics in the mixed sciences such as mechanics, optics, and navigation, which he expressed so evocatively in the river metaphor of 1669. Three central considerations motivated him in this.

First, it was driven by his conviction, already expressed in his youth, that the sole aim of philosophy consists in improving the lives of people and that the mathematical sciences must necessarily play a decisive role in achieving this aim[7]. The utilitarian justification of philosophy was of course widely propagated in the seventeenth century and was for instance written into the charter of the Royal Society. But for Leibniz its importance stretched beyond general claims of mechanistic and experimental science to the explanatory core of his hypothesis. In effect, the extraordinary success of the mathematical sciences needed, in his view, to be explained metaphysically. No other thinker had succeeded in providing such an explanation before him.

Second, accommodating the mathematical sciences on a structural level was part of Leibniz's broader conciliatory approach, often descriptively couched in terms of divine economy, by means of which he sought to evince the highest possible degree of veracity of his own hypothesis, be it the *Hypothesis physica nova* of his youth or the *Monadology* of his philosophical maturity[8]. All philosophical and scientific tradition contained in his view something of value as part of the heritage of human culture and learning[9]. But for Leibniz it was more than simply taking account of philosophical and scientific tradition which he set out to achieve. As he writes on

---

[7] See Leibniz, *Hypothesis physica nova*, conclusio, A VI, 2, 257.

[8] See Leibniz, *Hypothesis physica nova*, § 59, A VI, 2, 255; Leibniz to Arnauld, [29 September/9 October 1687], A II, 2, 249–250; Leibniz, *Reponse aux reflexions contenues dans la seconde Edition du Dictionnaire Critique de M. Bayle*, Gerhardt 1875-90 (cited hereafter as 'GP') IV, 568.

[9] See Leibniz, *Nouveaux Essais* III, 9, § 9, A VI, 6, 336–337.

one occasion, "it is no small indication of the truth of our hypothesis that it harmonizes all"[10]. If central tenets of ancient learning and modern scientific thought could be comfortably accommodated in his philosophy, there could be no better evidence for its truth.

Third, Leibniz was convinced that mathematical minds were able to introduce rigour into metaphysics which would otherwise be lacking, particularly as concerns our fundamental knowledge of the natural world. At the same time, he rejected the use of *mos geometricus* or geometrical method in philosophy, as this is found for example in Baruch de Spinoza's *Ethics*, on the grounds that such an approach represented an unsuitable methodological crossing of disciplinary borders[11]. When Pierre Bayle suggested in his *Remarques sur le systeme de M. Leibniz* that mathematicians who get involved in philosophical matters do not achieve much, Leibniz retorted that, on the contrary, they necessarily achieve more than most philosophers, "because they are accustomed to reason with precision"[12].

It is not necessary here to rehearse the many instances in which mathematical reasoning informed Leibniz's philosophical investigations. Much of his work on universal character and combinatorics sought to achieve a more exact, that is to say semantically unambiguous determination of the world around us, which would lead ultimately to the growth of knowledge, by starting out from basic elements or simple propositions. It suffices to say that already in his *Dissertatio de arte combinatoria*, first published in1666, the basic concept of the mechanistic philosophy, that all larger things are composed of smaller ones, be they atoms or molecules, was seen as providing the metaphysical basis for the universal application of combinatorics, since the fundamental relation of the latter, that of whole to part, is seen as having a direct correspondent in reality[13]. Even after Leibniz had disassociated himself from atomism, to which he was at times attracted[14], and was reintroducing the concept of substantial form into his philosophy, that is to say around 1679, he still described combinatorics as "a kind of metaphysical geometry"[15]. What is important here is that Leibniz for profound metaphysical and theological reasons rejected the reduction of nature philosophically to a mathematical model, for this would entail among other things the absolute necessity of human action. Nonetheless, he consistently employed from the time of his early writings on the theory of motion onward

---

[10] Leibniz, *Hypothesis physica nova* § 59, A VI, 2, 252: "Hypothesis nostra non parvo veritatis indicio omnes conciliat." See Beeley 1996, pp. 223–234.

[11] See Leibniz, *Recommandation pour instituer la science générale*, A VI, 4, 705.

[12] Leibniz to Masson, after [1]/12 October, 1716, GP VI, 628: "Il pretend que les Mathematiciens qui se mêlent de Philosophie, n'y reussissent gueres: au lieu qu'il semble qu'ils devroient reussir le mieux, étant accoustumés à raisonner avec exactitude."

[13] See Leibniz, *Dissertatio de arte combinatoria* § 34, A VI, 1, 187; Leibniz to Gallois, end of 1672, A II, 1 (2006), 354; Leibniz, *De arte inveniendi combinatoriae*, A VI, 4, 332; Beeley 1999, p. 138.

[14] See Arthur 2003.

[15] Leibniz, *De arte inveniendi combinatoria*, A VI, 4, 332: "Combinatoria agit quodammodo de Entium configuratione, seu coordinatione, nullo respectu habito loci, est quasi Geometria Metaphysica."

a model of the core of nature—albeit one which underwent a radical transformation from quasi-materialism to idealism by means of the concept of force—which had its origins in essentially mathematical concepts. Conversely his model of nature was able to facilitate the harmonization of mathematics and nature to a degree not found among any of his contemporaries. This essentially mathematical core is the reason why we find so many references in his letters and papers to "arcana rerum"[16], "arcana naturae"[17], "interieur de la nature"[18], and the like.

3 Metaphysics of Discovery and Explanation

The relationship between mathematics and model of nature in Leibniz is complex, as it comprises three components which do not always sit comfortably alongside one another.

First, there are genuinely metaphysical concepts which he postulates as part of the fundamental architectonics, most readily apparent in the hypothesis that basic material structures are replicated or folded into infinity, but also evident in questions concerning the composition of motion. The monadological concept of the existence of worlds within worlds *ad infinitum* always took its source from theoretical deliberations on infinite divisibility of the continuum[19].

Second, there is a heuristic aspect to the interrelationship, because we are led to expect precisely the agreement between the explanatory model and scientific discovery, that is to say between the *explanans* and the *explanandum* which is ultimately evinced. Thus for example the law of continuity, which as Leibniz never ceases to tell us[20], he first introduced into the republic of letters, is able to exclude false theories of motion, such as those originally proposed by Descartes and later elaborated by some of his followers. At the same time, the law of continuity derives from the principle of reason and is thus already written into the very constitution of the created world[21].

Third, there is the explanatory component itself, in which Leibniz on the basis of his model of nature seeks to provide metaphysical reasons for discoveries in contemporary mathematical—and indeed biological—science, which are often expressed in terms of divine benevolence. These explanations are essentially grounded in his metaphysics, for example in the concept that God is a perfect geometer or

---

[16] Leibniz, *Rationale fidei catholicae*, A VI, 4, 2321. See Beeley 2004, pp. 29–30.

[17] Leibniz, *Praefatio ad libellum elementorum physicae*, A VI, 4, 2003; Leibniz to Carcavy, beginning of November 1671, A II, 1 (2006), 288.

[18] Leibniz to Oldenburg, end of October 1676?, A II, 1 (2006), 380.

[19] See Leibniz to Queen Sophie Charlotte, [20]/31 October 1705, GP VII, 560.

[20] See for example Leibniz, *Tentamen anagogicum*, GP VII, 279; Leibniz, *Justification du calcul des infinitesimales par celuy de l'algebre ordinaire*, Gerhardt 1849-63 (cited hereafter as 'GM') IV, 105; Leibniz to De Volder, 1699, GP II, 192.

[21] See Duchesneau 1994.

that the continuum constitutes the base of God's reason. Here, not only are meta-physical grounds given for the scientific discoveries themselves, but also—as a kind of divine payback—post hoc justification for the very hypothesis that is obtained.

The beginning of this interrelationship between mathematics and nature can be observed already in Leibniz's first philosophical system which emerged during his employment as privy councillor in Mainz in the early 1670s. The concept of point which he developed in his *Theoria motus abstracti* mirrored precisely the concept of conatus, thereby allowing a direct correlation between lines and motions, math-ematics and phoronomy. It is to this early work of Leibniz which I now turn.

4   Metaphysics and Mathematics in Leibniz's Early Philosophy

Leibniz presents his concept of point in *Theoria motus abstracti* through a series of negations by means of which he sets it apart from the defining characteristics laid down by Aristotle and Hobbes. Just as the Aristotelian conception of unextended indivisible is rejected, so, too, the diametrically opposing Hobbesian conception of an infinitely small extended divisible whose parts and quantity simply do not enter any calculation:

> A point is not that whose part is nothing, nor that whose part is not considered; rather, it is that whose extension is nothing or that whose parts are without distance, whose magnitude is inconsiderable, indeterminable, smaller than that whose relation to another sensible mag-nitude can be expressed except as infinity, smaller than that which can be given. Moreover this is the foundation of the method of Cavalieri, through which its truth is plainly demon-strated when certain rudiments or beginnings of lines and figures are imagined to be smaller than any giveable line or figure."[22]

By conceiving points in this way, Leibniz believed he would be able to overcome the labyrinth of the continuum, whose composition had vexed philosophers since ancient times. The paradoxes which arose when attempting to compose a continuum out of true indivisibles– as atomism had sought to do at least implicitly—vanished once the categorical distinction between point and extension was largely negat-ed[23]. Both now consisted of parts, the only difference being that the parts of point were considered to be "indistant" (partes indistantes). It was for this reason that the young Leibniz was convinced that he was also able to save Cavalieri's method for determining quadratures and cubatures from critics like the Austrian mathematician

---

[22] Leibniz, *Theoria motus abstracti*, fund. praed. § 5, A VI, 2, 265: "Punctum non est, cujus pars nulla est, nec cujus pars non consideratur; sed cujus extensio nulla est, seu cujus partes sunt in-distantes, cujus magnitudo est inconsiderabilis, inassignabilis, minor quam quae ratione, nisi in-finita ad aliam sensibilem exponi possit, minor quam quae dari potest: atque hoc est fundamentum Methodi Cavalerianae, quo ejus veritas evidenter demonstratur, ut cogitentur quaedam ut sic dicam rudimenta seu initia linearum figurarumque qualibet dabili minora."

[23] See Beeley 1996, pp. 235–261. This interpretation is accepted by Arthur, although he confus-ingly calls Leibniz's actually divided points "indivisibles". See Arthur 2009, pp. 12–17.

and astronomer Paul Guldin who had attacked it on account of the implied composition of extension from indivisibles[24]. Significantly, the characterization of the magnitude of point as being smaller than any magnitude which can be given negates the absolute and thus opens up the possibility of quantitative relations between points themselves.

In *Theoria motus abstracti*, Leibniz defines the phoronomic concept of conatus analogously to the concept of point as the beginning or end of motion. Just as point is ontologically the limit of a line, so conatus is ontologically the limit of a motion. But points fulfilled this function already within the Aristotelian tradition, where they are conceived as being true indivisibles. The decisive theoretical development which Leibniz proposes in his early philosophy is that through infinite replication points can be seen to compose lines, just as through infinite forward replication conatus can be seen to compose motion: "Conatus is to motion as point is to space, or as one to infinity, it is namely the beginning or end of motion."[25]

Conatus is therefore clearly distinguished from rest, which as Leibniz emphasizes does not stand in relation to motion as point does to space, but rather in the relation of nothing to the number one[26]. Leibniz always believed that there could be no true state of rest in nature[27]. Everything on his view is constantly in flux, even if imperceptibly so; conatus is thus something like an infinitely small motion, endowed with its own direction. This belief was theoretically founded in his conviction that motion cannot be rationally defined in terms of its contrary, and provided him with one of his strongest reasons for rejecting the laws of motion propounded by Huygens and Wren. According to Huygens's first law of motion, a hard body in motion colliding with an equally hard body at rest would lose all its motion and transfer that motion to the body hitherto at rest[28]. Leibniz explicitly negates the idea that a resting body can cause another body to lose its motion[29].

As already mentioned, Leibniz's early concept of point allowed him to postulate the existence of quantitative relationships on a non-quantitative level of a kind which to a remarkable extent anticipated the quantitative relationships he postulated in his subsequent work on the calculus. The most thorough exposition of his concept of point at this time is to be found in the section of his theory of abstract motion entitled "Predemonstrable foundations" (fundamenta praedemonstrabilia). Drawing on the third of these foundations, which denies either in space or body the existence

---

[24] See Mancosu 1996, pp. 50–56.

[25] Leibniz, *Theoria motus abstracti*, fund. praed. § 10, A VI, 2, 265: "Conatus est ad motum, ut punctum ad spatium, seu ut unum ad infinitum, est enim initium finisque motus."

[26] Leibniz, *Theoria motus abstracti*, fund. praed. § 6, A VI, 2, 265: "Quietis ad motum non est ratio quae puncti ad spatium, sed quae nullius ad unum."

[27] See Leibniz, *De rationibus motus* § 12, A VI, 2, 161; *De mundo praesenti*, A VI, 4, 1511; *Nouveaux Essais*, préface, A VI, 6, 53, 56.

[28] Huygens 1669, p. 927: "1. Si Corpori quiescenti duro aliud aequale Corpus durum occurrat, post contactum hoc quidem quiescet, quiescenti vero acquiretur eadem quae fuit in Impellente celeritas."

[29] See Leibniz, *De rationibus motus* § 12, A VI, 2, 161: "Quies nullius rei causa est, seu corpus quiescens alii corpori nec motum tribuit, nec quietem, nec directionem, nec velocitatem."

of a minimum, where minimum is understood as that which has neither part nor magnitude, he argues in the fourteenth foundation for the interdependency of his concepts of point, temporal moment and conatus:

> But whatever moves at all is not at any time in one place while it moves, not even in a certain instant or minimum time, since what moves in time strives in an instant or begins or ceases to move, that is, it changes its place. Nor is it necessary to say that to strive in a time smaller than any time which can be given is assuredly to be in a minimal space, for there can be no minimal part of time or else there would be a minimal part of space. Then what covers the length a line in time covers the length of a line smaller than any line which can be given or a point in a time smaller than any time which can be given; and in an absolutely minimal time an absolutely minimal part of space, such as which cannot be the case according to foundation 3.[30]

This interdependency of space, time, and motion allows Leibniz to develop a further principle, namely that one point can be larger than another just as one conatus can be larger than another, on the necessary assumption of a *tertium comparationis*: that each temporal instant is equal to another.

> No-one can easily deny the inequality of conatus, for this follows from the inequality of points. [...] Therefore in a given instant the stronger conatus covers more spacethan the weaker, but no conatus can pass through more than a point or a part of space smaller than can be expressed in one instant; otherwise it could pass through an infinite line in time. Therefore one point is larger than another.[31]

Implicitly, the concepts of point, instant, and conatus are understood in *Theoria motus abstracti* as being infinitely small in relation to their corresponding quantitatively extensive concepts of space, time, and motion. Already in Mainz, the infinite difference between point and line, instant and time, as well as between conatus and motion, precludes on Leibniz's view that the one can be expressed in terms of the other.

---

[30] Leibniz, *Theoria motus abstracti*, fund. praed. § 14, A VI, 2, 265–266: "Sed et omnino quicquid movetur non est unquam in uno loco dum movetur, ne instanti quidem seu tempore minimo; quia quod in tempore movetur, in instanti conatur, seu incipit desinitque moveri, id est locum mutare: nec refert dicere, quolibet tempore minore quam quod dari potest, conari, minimo vero esse in loco: non enim datur pars temporis minima, alioquin et spatii dabitur. Nam quod tempore absolvit lineam, tempore minore quam quod dari potest, absolvet lineam minorem quam quae dari potest seu punctum; et tempore absolute minimo partem spatii absolute minimam, qualis nulla est *per fund. 3*."

[31] Leibniz, *Theoria motus abstracti*, fund. praed. 18, A VI, 2, 266–267: "Conatu[u]m inaequalitatem nemo facile negaverit, sed inde sequitur inaequalitas punctorum. [...] Ergo instanti dato fortiori spatii absolvet, quam tardior, sed quilibet conatus non potest percurrere uno instanti plus quam punctum, seu partem spatii minorem, quam quae exponi potest; alioquin in tempore percurreret lineam infinitam: est ergo punctum puncto majus."

5 Infinity, Error and Utility

Even on the basis of his still largely rudimentary knowledge of mathematics, if we are to believe his own accounts, Leibniz draws conclusions concerning the infinite which anticipate to a remarkable degree later work where he has a much more sophisticated conceptual framework in mathematics at his disposal. This is particularly evident in his treatment of curves. Although he at times allows a metaphysical distinction between a circle and its inscribed or described polygon in his writings of the early 1670s, Leibniz crucially argues that curves can be treated in the way established in Archimedes, because the resulting error is "smaller than can be expressed by us"[32] or, as he also puts it, because the resulting error is "imperceptible"[33]. After outlining a series of special problems such as the construction of a cylinder from mere rectilinear bodies or the construction of circular motion from a number of rectilinear motions, which appear to require mechanical solutions, he writes in *Theoria motus abstracti*:

> Even if these and other problems cannot be solved through abstract concepts of motion in bodies considered absolutely, they can nonetheless easily be explained in sensible bodies, namely under the assumption of an insensible ether, on account of which no sensible error disturbs our reasons, which suffices to explain phenomena. For sure, nature [...] and art solve these problems quite differently from the geometer, namely mechanically by means of motions that are not continuous but actually interrupted; just as when geometers describe the quadratrix by points, and Archimedes the circle by polygons, through the removal of error nothing will disturb the phenomena.[34]

As this passage suggests, Leibniz occasionally in his early work allows phenomenal considerations to run into theoretical considerations when talking about what might be called infinitely small quantities. But we should not allow this apparent conflation to obscure the importance of his remarks, formulated as they are, some 2 years before the beginning of his momentous stay in Paris. Leibniz was already clear at this time that only the existence of an imperceptible error could explain the applicability of mathematical reasoning to our understanding of natural phenomena. In this sense there must according to his metaphysics be an insignificant difference between what is mathematically exact and that which is natural and therefore

---

[32] Leibniz, *Pacidius philalethi*, A VI, 3, 569, var. (reconstructed): " Quemadmodum pólygonum regulare infinitorum laterum pro circulo metaphysice haberi non potest, tametsi in Geometria pro circulo habeatur, ob errorem minorem quam ut a nobis exprimi possit." Cf. Leibniz, *De infinite parvis*, A VI, 3, 434.

[33] Leibniz to Jean Chapelain (?), first half of 1670, A II, 1 (2006), 87; Leibniz, *Theoria motus abstracti*, usus, A VI, 2, 273; *De rebus in scientia mathematica tractandis*, A VI, 4, 379; Beeley 1999, p. 140.

[34] Leibniz, *Theoria motus abstracti*, usus, A VI, 2, 273: "Etsi haec aliave solvi non possent ex abstractis motus rationibus in corporibus absolute consideratis; in sensibilibus tamen, assumto saltem Aethere insensibili, facile explicari potest qua ratione efficiatur ut nullus error sensibilis rationes nostras turbet, quod phaenomenis sufficit. Nimirum longe aliter Natura [...] et Ars haec problemata solvit, quam Geometra; mechanice scilicet, motibus non continuis, sed revera interruptis; uti Geometriae describunt quadratricem per puncta, et Archimedes quadrat Circulum per Polygona, spreto errore nihil phaenomena turbaturo."

somewhat less than exact. Precisely this kind of approximation came into play when working with the infinite in mathematics. In a letter to Jean Chapelain, written in the first half of 1670, he reveals the results of his latest deliberations. Claiming that no curvilinear figure is imaginable which cannot be rationally expressed, he suggests that any error contained in the expression can be made less than what can be perceived. And this, he argues, is sufficient for practical purposes:

> In general, I maintain, as exemplified by the Elements of Euclid, that no figure can be contrived which cannot be explicated according to a prescribed method (if one does not mind subjecting oneself to laborious operations such as Archimedes did in the measurement of the circle, and also Ludolph van Ceulen, who has proceeded much further) in such a way that no error can be perceived, which is sufficient in practice.[35]

The fundamental significance of these remarks is clear: when working with the infinite, procedures can be pursued indefinitely; we can in principle increase the level of accuracy as much as we want. Almost incidentally, Leibniz refers to the long-windedness of the method of exhaustion employed by Archimedes in his work on the dimension of the circle, and suggests that Ludolph van Ceulen had substantially furthered our knowledge through his calculation of pi in *Van den Circkel* (1596). As is well known, Leibniz himself began work on determining the best possible approximation to pi already during his stay in Paris, although he did not publish details of these investigations until 1682.

The conflation of phenomenal and theoretical considerations during the Mainz period is not as extraordinary as might at first appear. Just as mathematics mirrors physics in respect of the elemental quanta point and conatus, so, too, does it do this in respect of procedures involving the infinite. When Leibniz sets out his model of nature in the *Hypothesis physica nova*, he postulates the existence of an infinitely replicated structure of bubbles (bullae) which combines the fundamental ideas of corpuscular theory with the infinite divisibility of matter. Nor does such a structure preclude the possibility of causal explanation. When considering the eventual causal contribution of events on a lower level to those on a higher level, the young philosopher suggests that they are of no consequence because they do not affect our phenomena. If there are worlds within worlds into infinity, as the micrographic investigations of men, such as Robert Hooke and Marcello Malpighi implied, they were for Leibniz at this time largely autonomous worlds from a causal point of view:

> And although according to the observations of micrographers there are continuously some things smaller than others, the same relations will always obtain. Since aqueous bubbles compared to air bubbles are like earth bubbles and air bubbles compared to ether bubbles have the same relation, nothing prevents the possibility of there being another ether, of which we have no suspicion, and which is loftier than the ether we have recognized through reasoning and experiment to the same extent that water is loftier than earth and air is loftier

---

[35] Leibniz to Jean Chapelain (?), first half of 1670, A II, 1 (2006), 87: "Omnino inquam, quemadmodum enim post extantia Euclidis Elementa, nulla est excogitabilis figura, quae non praescripta methodo (si quem non taedeat diuturnis subjectionibus operam impendere, uti in dimensione Circuli fecit Archimedes, et qui multo longius progressus est, Ludolphus a Colonia) solvi ita possit, ut error sit insensibilis, quod in praxi sufficit."

than water. But this ether cannot enter into our calculation, because on its account no phenomena are changed.[36]

The mathematization of nature in the young Leibniz effectively goes so far as to allow the identification of phenomenal and mathematical concepts of error. This is reflected in his assertion that when geometers describe the quadratrix by means of points, just as when Archimedes squares the circle by means of polygons, the resulting error is so small that phenomena are not disturbed[37]. And likewise it finds expression in the insensible ether, whose existence he postulates as an all-encompassing medium in the philosophical system of the *Hypothesis physica nova*. On account of this concept Leibniz is able to reconcile phenomena and theory in a thoroughly rational way, so that "no sensible error disturbs our reasoning, which satisfies the phenomena as well"[38].

6 Natural Phenomena and the Question of Exactness

The ancient concept of satisfying phenomena is central to the strategy Leibniz pursues in the mathematization of nature in his early writings. Decisively he distinguishes how things are in themselves from how they are perceived or how they are *ad sensum*—a distinction which he drops in his mature metaphysics on account of the fundamentally idealist tenet on which much of the theory of monads rests, namely that perception and expression are convertible. Tacitly presupposing the metaphysical principles which preclude mathematical necessity from entering his model of nature, Leibniz asserts in his *Hypothesis physica nova* that no perfectly uniform and continuous curvilinear or straight motion can occur in the natural world. Although there might appear to be such motion, this is, he says, only *ad sensum*, that is to say, it is only apparently the case.[39]

Likewise, the paths of falling bodies might appear parallel, but they are on his view actually curved, as a result of the combined circular motion of the earth in one direction and of ether in the other, as he sets out in the explanatory model on which

---

[36] Leibniz, *Hypothesis physica nova* § 49, A VI, 2, 243: "Et quamvis ex Micrographorum observationibus dentur continuo aliae aliis minores, manebit tamen semper eadem proportio: cum aqueae aëreis comparatae sint terreae, et aëreae ad aethereas eandem proportionem habeant, et nihil prohibeat dari alium aetherem, de quo nobis nec suspicari licet, aethere illo quem ratione et experimentis colligimus tanto superiorem, quanto est aqua terra, aut aër aqua. Sed haec in computum nostrum, quia nihil inde Phaenomena variantur, venire non possunt."

[37] Leibniz, *Theoria motus abstracti*, usus, A VI, 2, 273: "[…] uti Geometrae describunt quadratricem per puncta, et Archimedes quadrat Circulum per Polygona, spreto errore nihil phaenomena turbaturo."

[38] Leibniz, *Theoria motus abstracti*, usus, A VI, 2, 273: "Etsi haec aliave solvi non possent ex abstractis motus rationibus in corporibus absolute consideratis, in sensibilibus tamen, assumto saltem Aethere insensibili, facile explicari potest, qua ratione efficiatur, ut nullus error sensibilis rationes nostras turbet, quod phaenomenis sufficit."

[39] Leibniz, *Hypothesis physica nova* § 59, A VI, 2, 255.

his hypothesis rests[40]. Nonetheless, the absence of what counts as being mathematically exact in no way hinders the successful application of mathematics to our understanding of nature. While movements which appear straight are in themselves, that is to say on account of the contingencies of the existing natural world, actually curved, the curvature is, Leibniz suggests, so insensibly small, that all phenomena occur as if they were truly straight. The same applies to other phenomena, too: the flexibility and perspicuity of material bodies, the reflection and refraction of light, the behaviour of sound, and so on. The architectonics of his metaphysical model, as has already been noted, both explains and is confirmed by the successful application of mathematics. Mixed sciences drawn from physics and mathematics can be applied, since through the benevolence of God the phenomena in question not only appear to be of the highest exactitude, but also as far as our usage is concerned everything happens as if it were so:

> And here it is right that the geometrical practise of God in the economy of things be admired. For, although it is impossible in the nature of things that any physical body whatsoever be completely luminous, perspicuous, fluid, firm, soft, stretchable, bendable, hard, warm, and so on; or that motion be exactly continuous, uniform, uniform increased or decreased, rectilinear, circular, reflected, refracted, changed; or that the effect of a magnet, of light and sound reach any assignable point, and so on, all this happens nonetheless to the highest degree of exactness for the senses: although they are not, they nonetheless appear to the senses to be so. And as far as our usage is concerned, it is as if they were so; and moreover through an extraordinary good deed of God optics, music, statics, elastics, the doctrine of impacts (or concerning impetus and percussion), myology or concerning the motion of muscles, and even pyrotechnics and general mechanics, and whatever is a combination of physical and mathematical sciences, can be perfected in theorems which to the envy of the pure sciences, do not fail to satisfy the senses.[41]

It is important to recognize that the systematic role which Leibniz accords to the mathematical sciences in our understanding of nature corresponds in a strict sense to his views on the ontological status of mathematical objects themselves. Already

---

[40] Leibniz, *De firmitate, vi elastica, explosione, attractione*, A VI, 4, 2082: "[...] uti ad sensum gravium directiones apud nos sunt parallelae."; Leibniz to Queen Sophie Charlotte, [20]/31 October 1705, GP VII, 563: "Mais la Nature ne peut point, et la sagesse divine ne veut point tracer exactement ces figures d'essence bornée, qui presupposent quelque chose de determiné, et par consequent d'imparfait, dans les ouvrages de Dieu. Cependant elles se trouvent dans les phenomenes ou dans les objets des esprits bornés: nos sens ne remarquent point, et nostre entendement dissimule une infinité de petites inegalités qui n'empêchent pourtant pas la parfaite regularité de l'ouvrage de Dieu, quoyque une creature finie ne la puisse point comprendre." See also Beeley 1995.

[41] Leibniz, *Hypothesis physica nova* § 59, A VI, 2, 255: "Atque hic admirari licet praxin DEI in oeconomia rerum geometrisantis. Etsi enim per naturam rerum impossibile sit, corpus aliquod totum lucere, perspicuum, fluidum, grave, molle, tendibile, flexibile, durum, calidum etc., item motum continuum, uniformem, uniformiter acceleratum vel diminutum, rectilineum, circularem, reflexum, refractum, permutatum, exacte esse; effectum magnetis, luminis et soni ad quodlibet punctum assignabile pervenire etc. Evenit tamen ut summa ad sensum ἀκριβείᾳ haec omnia, etsi non sint ita, tamen sensu esse videantur, ut quantum ad usum nostrum, perinde sit ac si essent; atque ita incredibili Dei beneficio, Optica, Musica, Statica, Elastica, πληγικὴ (seu de impetu et percussione), Myologia seu de motu musculorum, imo et Pyrotechnica et Mechanica universa, et quidquid est mixtarum ex Physica Mathematicaque scientiarum, ad purarum invidiam usque, non fallentibus ad sensum (nisi per accidens) theorematibus excoli possint."; Beeley 1999, p. 132.

in the first draft of his *Theoria motus abstracti*, he deliberated on the nature of geometrical and physical construction and found that no-one up to that time had propounded the true, physical, real and exact causes of figures in the world, "for the causes of the geometers are imaginary"[42]. He then cites as an example the exact construction of a sphere from mere straight lines. Such deliberations are widespread in his writings from the 1670s in general and a number of particularly noteworthy instances are to be found in some of the papers on the arithmetical quadrature of the circle which will soon be published in the Academy Edition for the first time[43].

On the evidence of numerous manuscript drafts which Leibniz produced in 1677, it seems that he was fairly clear by that time that mathematics carries its proof in itself[44] and therefore does not refer essentially in any way to things in nature. In a piece to which he aptly gave the title *Dialogus,* Leibniz's fictional opposite number points out that "when we inspect geometrical figures we often elicit truths through their accurate consideration". To this Leibniz replies:

> It is so, but it must be recognized that these figures are treated as characters, nor is for instance a circle described on paper a true circle, nor is it necessary, rather, it suffices that it be treated by us as a circle.[45]

Correspondingly, as he remarks in the contemporary piece entitled *La vraie methode*, referring among others to van Ceulen's calculation of pi to twenty decimal places, he argues that proofs set out on paper in order to avoid false reasoning "are not made in the thing itself, but on the characters which we have substituted in place of these things"[46]. Even before this time he had established that the truth of a mathematical proof which rests on a geometrical construction does not lie in the actual construction itself. This could not be the case, Leibniz argued, because all visual mathematical objects brought about by artificial means—he generally calls these mechanical constructions—are necessarily inexact. Falling back on established tradition he contrasts these with geometrical constructions, which contain procedures

---

[42] Leibniz, *Theoria motus abstracti*, first draft, A VI, 2, 184: "Et certe nemo hucusque tradidit causas versa, physicas, reales, exactas figurarum, quibus eas in mundo produci necesse est, nam causae geometrarum sunt imaginariae, v.g. quomodo non mechanice, sed exacte sphaera ex meris rectilineis fieri possit, nulla alia sphaera, imo curvilineo nullo, praesente."

[43] See for example the piece entitled *Dissertationis de arithmetica circuli quadratura propositiones septem*, dated early 1674, to be published in A VII, 6.

[44] See Leibniz, *La vraie methode*, A VI, 4, 4: "Cette raison est, que les Mathematiques portent leur épreuve avec elles."

[45] Leibniz, *Dialogus*, A VI, 4, 23: "B. At quando figuras Geometriae inspicimus saepe ex accurata earum meditatione veritates eruimus. A. Ita est, sed sciendum etiam has figuras habendas pro characteribus, neque enim circulus in charta descriptus verus est circulus, neque id opus est, sed sufficit eum a nobis pro circulo haberi." See also Leibniz, *De veritatis realitate* (1677), A VI, 4, 18: "Etiamsi nullus existat reapse circulus."

[46] Leibniz, *La vraie methode*, A VI, 4, 5: "Il faut donc remarquer que les preuves ou experiences qu'on fait en mathematique pour se garantir d'un faux raisonnement (comme sont par exemple la preuve par l'abjection novenaire, le calcul de Ludolph de Cologne touchant la grandeur du cercle; les tables des sinus ou autres) ne se font pas sur la chose même, mais sur les caracteres que nous avons substitués à la place de la chose."

by which the production of a figure is possible, whereby the veracity of the figure is guaranteed when the cognitive principle of non-contradiction is upheld. Geometrical constructions are thus already for the young Leibniz "imaginary but exact", since they contain procedures "according to which bodies can be constructed, although often by God alone."[47] It is this ideal status of mathematical objects which informs Leibniz's metaphysics from beginning to end, precisely as the river metaphor formulated in 1669 already makes clear.

7 Mathematics and the Improvement of Human Understanding

The ontological ideality of mathematical objects is seen by Leibniz as an essential part of its usefulness as a discipline in at least one crucial respect: that through its exercise mathematics can on his view contribute directly to increasing the perfection of human understanding. Indeed, on numerous occasions he describes this as being one of the main uses of mathematics alongside that of contributing to improvements to the life of the general public. When he writes to the influential French minister of finance, Colbert, at the end of 1679, he emphasizes the importance of mathematical education as a means to promoting the public good:

> When we now consider the innermost part of geometry, we stay in the cortex of nature. Natural science is more useful to our body; geometry is more efficacious in perfecting the mind. Whence it is not surprising, how much stronger geometry is pleasing to the intellect, even in natural science itself. Archimedes, while he invented so many other useful things, detected that the sphere and cylinders are related among themselves, and ratified this by drawings in a small mound of sand. It is important that geometry is not neglected in the state, for it sharpens intellects and teaches strict reasoning.[48]

France was always associated in Leibniz's mind with his own mathematical education. In a letter to the Helmstedt professor of law, Hermann Conring, written shortly after his arrival in Hanover, Leibniz describes natural philosophy as being "nothing else than concrete mathematics or mathematics exercised in matter, just as in optics

---

[47] Leibniz, *Theoria motus abstracti*, probl. gen., A VI, 2, 270: "Geometrica continet modos, quibus corpora construi possunt, licet saepe a solo Deo."

[48] Leibniz to Jean Baptiste Colbert, December 1679, A III, 2, 919: "Deinde Geometriae intima pervidemus, in naturae cortice haeremus. Utilior corpori nostro physica: ad perfectionem intellectus efficacior Geometria. Unde mirum non est, quanto quisque ingenio validior Geometria magis, etiam in ipsa physica, delectari. Archimedes cum tot alia utiliora vitae invenisset, sphaeram tantum et cylindorum quorum inter se relationem deprehenderat, tumulo insculpi jussit. Geometriam non negligi Reipublicae interest, nam ingenia acuit, et severe ratiocinari docet."; Leibniz to Friedrich Wilhelm Bierling, [26 June]/7 July 1711, GP VII, 496, Leibniz to Hermann Conring, 3/[13] January 1678, II, 1 (2006), 581. See also the piece entitled *Introductio ad praefationem libelli geometrici*, to be published in A VII, 6: "Duplex est Geometriae utilitas, nam vel ad vitae praesentis commoditates pertinet vel ad ipsam per se mentis perfectionem refertur."

or music"[49], before he proceeds to describe the training in firm and certain thought which he had experienced during his stay in Paris. As evidence of the successful results of the diligence and time he had devoted to the study of geometry, he cites his work on an infinite series for expressing pi, pointing out that he could in fact give many other examples of his mathematical work although none of these had been published. Significantly, when discussing the series, which today bears his name and which in modern terms can be expressed thus

$$\sum_{k=0}^{\infty} (-1)^k/(2k+1) = 1 - 1/3 + 1/5 - 1/7 + 1/9 \ldots = \pi/4$$

he emphasizes, as we have already seen, not only that the series can be continued at will into infinity, but also that in thus proceeding the error will become less than any given quantity. Already in his pre-Paris writings, Leibniz had recognized that this guarantees not only certainty, but also usage.

Two things are important in this example. First, it is no coincidence that he moves from discussing the mental training afforded by mathematical—and one might add logical—study to an example of his work on the quadrature of the circle, for he was at pains particularly at this time to emphasize that despite the finiteness of the human mind we are able to know much about the infinite. In fact, he stresses that we can know the infinite, not that we can understand it. At the end of the 1670s Leibniz was in the process of developing a highly sophisticated philosophical concept of infinity, partly based on his own work on quadratures and infinite series, partly drawing on the work of others such as Evangelista Torricelli and John Wallis on calculating the volumes of infinitely long solids.[50]

Second, it reflects his continuing interest in the concept of error, both in the sense of negligible error when working with infinite procedures in mathematics, and in the sense of errors occurring in mathematical reasoning—which as we have already seen pertains directly to the ontological status of mathematical objects. Since,

---

[49] Leibniz to Hermann Conring, 3/[13] January 1678, A II, 1 (2006), 581–582: "Geometriae enim usus in applicatione consistit, abstracta autem tantum ingenio exercendo et solidis certisque assuescendo servit, quod me in Gallia fecisse, majore fortasse animi contentione quam necesse erat, nunquam poenitebit. Nam ab eo tempore quo diligentius Geometriae dedi operam, de rebus omnibus paulo curatius judicare coepi. Specimina studii mei Geometrici habeo multa, sed nondum edita; unum mirifice amicis in Gallia Angliaque placuit, quod demonstravi, Quadratum circumscriptum Circulo esse ad ipsum Circulum ut 1/1 ad 1/1–1/3 + 1/5–1/7 + 1/9–1/11 + 1/13–1/15 etc. in infinitum, id est posito quadrato unitate, Circulo magnitudinem in numeris rationalibus simplicissime exprimi per hanc seriem seu harum fractionum alternatim additarum et subtractarum aggregatum. Eadem series servit ad appropinquandum pro lubitu in infinitum. Nam 1/1 est quantitas major circulo, sed error est minor 1/3, at 1/1–1/3 est quantitas minor circulo [...] et ita procedi potest in infinitum, ita ut error possit reddi minor quolibet assignato." See also Leibniz to François de la Chaise, second half of April/first half of May 1680, A III, 3, 191; Leibniz to Gerhard Molanus for Arnold Eckhard, beginning of April 1677, A II, 1 (2006), 281–282.

[50] See Leibniz 1993, p. 36, 104, 132 (*De quadratura arithmetica circuli*, prop. VII, schol., prop. XLV schol., variant to prop. XI schol.); Mancosu and Vailati 1991.

as Leibniz makes clear the ideality of mathematical objects in themselves also stretches to mathematical proofs—demonstrative reasoning, he writes to Sophie, Dowager Electress of Hanover in June 1700 "is based on inner light independent of sense"[51]—errors in human reasoning are primarily to be explained on his view through lack of attention of bad memory[52]. Error understood in this way is in us and quite definitely not in the object. Precisely with a view to perfecting the mind and achieving the cognitive rigour associated with this, Leibniz wrote in March of the previous year to Nicolas Malebranche that "mathematicians have just as much need to be philosophers as philosophers to be mathematicians"[53]. The central consideration thereby, as he had set out years earlier in the *Recommandation pour instituer la science générale*, is that geometers have many means of discovering the smallest errors, even if through inattention or distraction these should escape them. In philosophy, by contrast, it is necessary to employ precisely that exact reasoning which one finds in geometry, "because here other means of assuring oneself of correctness are most often missing"[54].

8 Deviation of the Infinite Series

When Leibniz described natural philosophy to Conring as "mathematics exercised in matter", he was expressing an idea which had informed the architectonics of his diverse metaphysical models of nature since 1666. Although he was consistently of the view that a philosophically tenable account cannot be comprehensively reduced to mathematical concepts, as Descartes and Hobbes had thought, he was nonetheless convinced that metaphysical reality can be grasped by mathematical concepts much in the way that a curve approximates to its asymptote—indeed, this is what he means when he occasionally in this context talks about "degrees" or "perfections" of reality[55]. The eternal truths of mathematics are therefore in a very profound sense

---

[51] Leibniz to the Electress Sophie, June 1700, GP VII, 553: "mais on ne sera jamais asseuré de la necessité de la chose sans appeller à son secours les raissonnemens demonstratifs, fondés sur la lumiere interne independante des sens."

[52] See Leibniz, *Animadversiones in partem generalem Principiorum Cartesianorum*, I §§ 31, 35, GP IV, 361. See also Leibniz to Queen Sophie Charlotte, mid-June 1702, GP VI, 501–502; *Extrait d'une letter de M. D. L. sur son Hypothese de philosophie*, GP IV, 501.

[53] Leibniz to Nicolas Malebranche, 13/23 March 1699, GP I, 356: "Les Mathematiciens ont autant besoin d'estre philosophes que les philosophes d'estre Mathematiciens."

[54] Leibniz, *Recommandation pour instituer le science générale*, A VI, 4, 705: "C'est dans la philosophie qu'il faudroit employer principalement cette rigeur exacte du raisonnement parce que les autres moyens de s'asseurer y manquent le plus souvent. "

[55] See Leibniz to de Volder, [26 August]/6 September 1700, GP II, 213: "Quae omnia jam dudum innueram, ne thesin prima fronte a plerisque admittendam precario prorsus assumsisse viderer, eaque rursus attingo non renovandi priora studio, sed ut pulcherrimae rei fontes intimius cognoscantur consteteque principia naturae non minus metaphysica quam mathematica esse, vel potius causas rerum latere in metaphysica quadam mathesi, quae aestimat perfectiones seu gradus realitatum."

constitutive of nature; they can never be contradicted by phenomena, he writes in 1702 to Burchard de Volder, because the difference is always smaller than any magnitude which can be assigned[56]. Clearly, "difference" is here understood in the sense of mathematical procedures for dealing with negligible error. Drawing on his work in analysis, Leibniz calls the deviation between natural phenomena on the one side and mathematical expressions expressing these natural phenomena on the other side the "deviation of the infinite series"[57].

As has already been mentioned, Leibniz explicitly rejects the employment of geometrical method in philosophy and this is reflected in his evident, often repeated distinction between reason and appearance. In his letter to the Dowager Electress Sophie of 31 October 1705, he points out in similar fashion that while our senses do not notice the difference our understanding effectively hides "an infinity of small inequalities"[58]. Our senses are limited, we do not *perceive* small differences in things. Our minds, in contrast, *conceal* these small differences so as not to hinder application. Mathematics, "the masterpiece of human reasoning", can therefore serve to judge the truth of sensible things, because architectonically phenomena must conform to the laws of the mixed sciences through which mathematics is applied to nature. We say here that the phenomena *must* conform, precisely because the mind is partly involved in *making them* conform.

Often Leibniz's remarks emphasize the benefits which accrue to mankind through the mixed sciences. Thus he describes physics in a letter to Jacob Auguste Barnabas, Comte des Viviers as "divine mathematics exercised in nature" and suggests that the mathematical sciences will lead to discoveries "which could remedy our evils and make the political state perfect"[59]. Again, much of this is anticipated in Leibniz's pre-Paris writings. Already in the *Hypothesis physica nova*, he talks of theorems being perfected in the mixed sciences which, to the envy of the pure sciences cannot be failed by the senses except by accident, and in a similar way to in his later writings he puts this down to the incredible benevolence of God[60].

---

[56] Leibniz to de Volder, [8]/19 January 1702, GP II, 282–283: "Interim scientia continuorum hoc est possibilium continet aeternas veritates, quae ab actualibus phaenomenis nunquam violantur, cum differentia semper sit minor quavis assignabili data." See Knobloch 2008, pp. 176–177.

[57] Leibniz, *Specimen inventorum de admirandis naturae generalis arcanis*, A VI, 4, 1622: "Nam ex eo quod nullum corpus tam exiguum est, quin in partes diversis motibus incitatas actu sit divisum, sequitur nullam ulli corpori figuram determinatam assignari posse, neque exactam lineam rectam, aut circulum, aut aliam figuram assignabilem cujusquam corporis reperiri in natura rerum, tametsi in ipsa seriei infinitae deviatione regulae quaedam a natura serventur."

[58] Leibniz to Queen Sophie Charlotte, [20]/31 October 1705, GP VII, 563: "nos sens ne remarquent point, et nostre entendement dissimule une infinité de petites inegalités qui n'empêchent pourtant pas la parfaite regularité de l'ouvrage de Dieu, quoyque une creature finie ne la puisse point comprendre."

[59] Leibniz to Des Viviers, May 1692, A I, 8, 270: "Ainsi nous pourrons tourner tout la force de nos esprits à la physique, c'est-à-dire à la Mathematique divine exercée dans la nature, et à la decouverte de ce qui pourroit remedier à nos maux et perfectionner nostre estat."

[60] Leibniz, *Hypothesis physica nova* § 59, A VI, 2, 255: "Evenit tamen ut summa ad sensum ἀκριβείᾳ haec omnia, etsi non sint ita, tamen sensu esse videantur, ut quantum ad usum nostrum, perinde sit ac si essent; atque ita incredibili Dei beneficio."

The conceptual basis for this beneficial application of the mixed sciences lies decisively in the concept of negligible error. Leibniz does not tire of telling us that there is nothing in nature which exactly corresponds to the notions of mathematics. Just as there are no perfectly curvilinear motions of things, so, too, are no mathematically precise figures are to be found in things; no true circles, no true ellipses, no true straight lines, and so on. Such figures are ideal and are as such only definable in the intellect[61]. Partly for the same reason there can in his opinion also be no perfect similitude in nature, for this would not only presuppose the natural existence of an essentially mathematical concept, but would in addition have the metaphysically harmful—when not to say with a view to the architectonics untenable—consequence of destroying individuality. Mathematical concepts are for Leibniz necessarily incomplete or underdetermined and they cannot therefore be reified[62]. Crucially on Leibniz's view the mixed sciences succeed in explaining natural phenomena, because the deviation of these phenomena from the normative values of mathematics is so small that the error in calculation which results is smaller that any given error. And this he tells us, drawing on his mathematical practice since Paris, is sufficient in order to demonstrate certainty as well as usage[63]. He makes the source of these deliberations eminently clear in a letter which he wrote to the Dowager Electress Sophie of Hanover on 31 October 1705:

---

[61] See for example *Hypothesis physica nova* § 59, A VI, 2, 255: "Etsi enim per naturam rerum impossibile sit, corpus aliquod totum lucere, perspicuum, fluidum, grave, molle, tendibile, flexibile, durum, calidum etc., item motum continuum, uniformem, uniformiter acceleratum vel diminutum, rectilineum circularem, reflexum, refractum, permutatum, exacte esse; effectum magnetis, luminis et soni, ad quodlibet punctum assignabile pervenire, etc. Evenit tamen ut summa ad sensum akribeia haec omnia, etsi non sint ita, tamen sensu esse videantur, et quantum ad usum nostrum, perinde sit ac si essent." Similar remarks are found in later pieces: *Mira de natura substantiae corporae*, A VI, 4, 1465: "[...] revera nullae certae figurae extant in natura rerum ac proinde ne certi motus"; *Dans les corps il n'y a point de figure parfaite*, A VI, 4, 1613: "Il n'y a point de figure precise et arrestée dans les corps à cause de la division actuelle des parties à l'infini"; *Principia logico-metaphysica*, A VI, 4, 1648: "Non datur ulla in rebus actualis figura determinata, nulla enim infinitis impressionibus satisfacere potest. Itaque nec circulus, nec ellipsis, nec alia datur linea a nobis definibilis nisi intellectu, ut lineae antequam ducantur, aut partes antequam abscindantur"; *Specimen inventorum de admirandis naturae generalis arcanis*, A VI, 4, 1622: "Nam ex eo quod nullum corpus tam exiguum est, quin in partes diversis motibus incitatas actu sit divisum, sequitur nullam ulli corpori figuram determinatam assignari posse, neque exactam lineam rectam, aut circulum, aut aliam figuram assignabilem cujusquam corporis reperiri in natura rerum, tametsi in ipsa seriei infinitae deviatione regulae quaedam a natura serventur"; *Definitiones cogitationesque metaphysicae*, A VI, 4, 1400. See also *Elementa nova matheseos universalis*, A VI, 4, 514.

[62] It is the approximation to the normative values of mathematics rather than abstraction on the part of the knowing subject which is decisive for Leibniz's concept of the success of the mixed sciences. Cf. Rutherford 1995, pp. 88–89.

[63] Leibniz, *De organo sive arte magna cogitandi*, A VI, 4, 159: "Nam etiamsi non darentur in natura nec dari possent rectae ac circuli, sufficiet tamen dari posse figuras, quae a rectis et circularibus tam parum absint, ut error sit minor quolibet dato. Quod satis est ad certitudinem demonstrationis pariter et usus."

Nevertheless the eternal truths founded on the limited ideas of mathematics do not fail to serve us in practice, so long as it is allowed to disregard those errors which are too small to cause errors of consequence in relation to the aim which one proposes; just as an engineer who draws a regular polygon on the ground does not need to trouble himself if one side is longer than another by several inches.[64]

It is not necessary here to cite the many places in Leibniz's letters and papers where he develops his ideas on negligible errors when working with the infinite. Some of these places are more philosophical, some more mathematical in character. What is common to them is the emphasis that he puts on showing that neither certainty nor usefulness is compromised when using infinite procedures in mathematics. Particularly remarkable is that these ideas are to be found, admittedly in a more rudimentary manner, already in writings which preceded his largely autodidactic mathematical training in Paris. Moreover, the notion that the infinite is not only deeply embedded in nature's core—"nature bears everywhere the character of the infinite", he remarks on one occasion to Des Billettes[65] –, but also must necessarily provide the key to understanding how mathematics and nature meet in physics or in the mixed sciences is present in his philosophical thought well before he turns his mathematical mind to questions of the infinite. And this is because the concept of negligible error seems to have been substantially derived from the overarching principle of usefulness so important in early modern philosophy and science rather than precise mathematical considerations on infinite procedures which only later acquire prominence in his work. Of course, it is not unlikely that Leibniz already in Mainz became acquainted with the basic principles of the classical method of exhaustion, perhaps through reading mathematical and physical sections of Hobbes's works in Boineburg's library[66], but the conclusions which he draws before he has achieved a thorough grounding in mathematics are nonetheless remarkable—not least in view of their longevity in his thought.

---

[64] Leibniz to the Electress Sophie, [20]/31 October 1705, GP VII, 563–564: "Cependant les verités eternelles fondées sur les idées mathematiques bornées ne laissent pas de nous servir dans las practique, autant qu'il est permis de faire abstraction des inegalités trop petites pour pouvoir causer des erreurs considerables par rapport au but qu'on se propose; comme un ingenieur qui trace sur le terrain un polygone regulier ne se met pas en peine si un costé est plus long que l'autre de quelques pouces." See also Leibniz, *Reponse aux reflexions*, GP IV, 569.

[65] Leibniz to des Billettes 15/25 March 1697, A I, 13, 656: "La nature seroit peu de chose si elle estoit epuisable à des esprits finis, elle qui porte en tout le caractere infini. On a pour maxime que la nature abhorre l'infini, et c'est tout le contraire pour moy." Similarly, Leibniz writes to the Electress Sophie in respect of van Helmont on [23 September]/3 October 1694, A I, 10, 61: "J'approuve sur tout son sentiment de l'infinité des choses, et j'ay deja dit dans le journal de sçavans que chaque partie ayant des parties, à l'infini, il n'y a point de petite portion de la matière qui ne contienne une infinité actuelle de creatures et apparement de creatures vivantes. C'est par là que la nature porte par tout le caractere de son Creature." See also *Considerations sur la différence qu'il y a entre l'analyse ordinaire et le nouveau calcul des transcendantes*, GM V, 308.

[66] On Leibniz's reception of Hobbes during his time in Mainz see Goldenbaum 2008, pp. 67–76.

9   The Justification of Infinitesimals

Usefulness was a criterion which Leibniz regularly adduced when seeking to justify the foundations of his infinitesimal calculus in the face of considerable criticism levelled against it. This is true already of his rigorously conceived infinitely small or infinitesimal quantities, which he characterizes as mental fictions comparable to the imaginary roots used in algebra[67], and with which he believes he is able to achieve something approaching direct proof in quadratures[68]. Part of the justification for these quantities was that in this way the circuitous apagogical proofs of Archimedes could be avoided[69]. Indeed, he claims on occasion that the concept of the infinitesimal in the sense of disappearing or evanescent quantities is not only useful but also founded in reality, since such quantities occur in the phenomena of natural change and motion.

This of course opens up the whole question of whether Leibniz really held that infinitesimals could exist in nature. On some occasions he does indeed seem to be denying their existence. But I think that we need to be careful here, because denial of the existence of infinitesimals is generally coupled with the argument that the success of the calculus does not depend on metaphysical discussions concerning reality. When he makes such claims, this seems to be no more than a get-out-clause vis-à-vis opponents who seek to provide metaphysical arguments against his calculus. Seen within the context of Leibniz' dynamics, particularly in respect of dead force (vis mortua) it is evident that he must be committed in some way to the existence of infinitesimals[70]. The emphasis which Leibniz places on using these quantities is precisely on account of their practicality in shortening proofs which would otherwise have to be much more elaborate and because at the same time this can be done without compromising rigour. Metaphysical disputes on their nature have, he argues, no weight when measured against established criteria of rigorous

---

[67] Leibniz to Johann Bernoulli, [28 May]/7 June 1698, GM III, 499: "Fortasse infinita, quae concipimus, et infinite parva imaginaria sunt, sed apta ad determinanda realia, ut radices quoque imaginariae facere solent."; Leibniz to Varignon, [3]/14 May 1702, GM IV, 98.

[68] See for example Leibniz, *Observatio quod rationes sive proportiones non habeant locum circa quantitates nihilo minores, et de vero sensu methodi infinitesimalis*, GM V, 389; Leibniz to Varignon, [22 January]/2 February 1702, GM IV, 92; Leibniz to Des Bosses [13]/24 January 1713, GP II, 305.

[69] Leibniz, *Historia et origo calculi differentialis*, GM V, 410: "Et dum ille rationes nascentes aut evanescentes considerat, prorsus a differentiali calculo abduxit ad methodum exhaustionum, quae longe diversa est (etsi suas quoque utilitates habeat) nec per infinite parvas, sed ordinarias procedit, etsi in illis desinat."; Leibniz, *Responsio ad nonnullas difficultates a Dn. Bernardo Niewentiit circa methodum differentialem seu infinitesimalem motas*, GM V, 322; Leibniz to Pinsson, [18]/29 August 1701, A I, 20, 494: "Car au lieu de l'infini ou de l'infiniment petit, on prend des quantités aussi grandes et aussi petites qu'il faut pour que l'erreur soit moindre que l'erreur donnée. De sorte qu'on ne differe du style d'Archimede que dans les expressions qui sont plus directes dans nostre Methode, et plus conformes à l'art d'inventer."

[70] This conclusion has, however, recently been rejected by Garber, who suggests that dead force 'should not be identified with its mathematical representation, and the reality of dead force should not be taken to entail the reality of infinitesimals'. See Garber 2008, pp. 303–304.

mathematical proof[71]. Nonetheless, as is well known, he takes these disputes seriously and at once weakens their force by allowing that infinitesimals be substituted by incomparables, that is to say, quantities which can be chosen smaller than any quantities which can be given[72]. In this way Leibniz claimed that he was able to reformulate any proof involving infinitesimals into a proof in the style of Archimedes using the method of exhaustion, albeit a considerably shortened version of that method. Thus in a letter to François Pinsson, part of which was published under the title *Mémoire de Mr. G. G. Leibniz touchant son sentiment sur le calcul differentiel, in the Journal de Trévoux* in 1701, Leibniz states:

> There is no need to take the infinite in a rigorous way, but only in the way in which one says in optics that the rays of the sun come from an infinitely distant point and are therefore taken to be parallel. And when there are several degrees of infinity, or infinitely small, this is like as when the globe of the earth is taken to be a point in comparison to the distance of the fixed stars, and a ball that we handle is still a point in comparison to the radius of the globe of the earth. So that the distance of the fixed stars is an infinitely infinite or infinite of the infinite in relation to the diameter of the ball. For, in place of the infinite or of the infinitely small, one can take quantities as great or as small as one needs so that the error be less than the given error. In this way one does not differ from Archimedes' style but for the expressions which in our method are more direct and more in accordance with the art of discovery.[73]

His true infinitesimals are more in accordance with the ars *inveniendi* because they open up new ways of reasoning, in particular, they allow us to avoid the cumbersome methods of the ancients.[74]

Already in the first major work presenting his new infinitesimal calculus, *De quadratura arithmetica circuli*, written in 1676, toward the end of his stay in Paris, Leibniz answers those who might question the economy of his method, the quality

---

[71] See Leibniz, *Reponse aux reflexions contenues dans la seconde edition du Dictionnaire de M. Bayle, article Rorarius, sur le systeme de l'harmonie préétablie,* GP IV, 569: "Les Mathématiciens cependant n'ont point besoin du tout des discussions métaphysiques, ni de s'embarrasser de l'existence réelle des points, des indivisibles, des infiniment petits, et des infinis à la rigueur."

[72] See Leibniz to Wallis, [20]/30 March 1699, GM IV, 63: "Verae interim an fictitiae sint quantitates inassignabiles, non disputo; sufficit servire ad compendium cogitandi, semperque mutato tantum stylo demonstrationem secum ferre; itaque notavi, si quis incomparabiliter vel quantum satis parva pro infinite parvis substituat, me non repugnare."

[73] Leibniz, *Mémoire de Mr. G. G. Leibniz touchant son sentiment sur le calcul différentiel,* GM V, 350: "[...] on n'a pas besoin de prendre l'infini ici à la rigueur, mais seulement comme lorsqu'on dit dans l'optique, que les rayons du Soleil viennent d'un point infiniment éloigné, et ainsi sont estimés parallèles. Et quand il y a plusieurs degrés d'infini ou infiniment petits, c'est comme le globe de la Terre est estimé un point à l'égard de la distance des fixes, et une boule que nous manions est encore un point en comparaison du semidiamétre du globe de la Terre, de sorte que la distance des fixes est un infiniment infini ou infini de l'infini par rapport au diamétre de la boule. Car au lieu de l'infini ou de l'infiniment petit, on prend des quantités aussi grandes et aussi petites qu'il faut pour que l'erreur soit moindre que l'erreur donnée, de sorte qu'on ne diffère du stile d'Archimède que dans les expressions, qui sont plus directes dans nôtre méthode et plus conformes à l'art d'inventer." (This text corresponds to that contained in Leibniz to Pinsson, [18]/29 August 1701, A I, 20, 493–494.) See also Leibniz to Varignon, [22 January]/2 February 1702, GM IV, 92; Leibniz, *Tentamen anagogicum. Essai anagogique dans la recherche des causes,* GP VII, 277; Knobloch 2008, pp. 176–177.

[74] See Knobloch 1999, pp. 215–216.

of his proof, by replying that the error which results from employing his infinitely small quantities is smaller than any error which can be given and is therefore null[75]. We do not need the solution so accurately, is the point he is putting across there, but simply what is sufficient *ad usum* vitae. The central concept is again that of applicability. He could, he suggests, in principle increase the accuracy, he could embellish the reasoning with extensive proofs. But accuracy beyond what can be shown to suffice in order for rigour to obtain does not suffice in itself, rather it is also and indeed decisively questions of applicability, usefulness[76].

Leibniz employs similar expressions then talking in general terms about the application of mathematics to nature in the mixed sciences. Despite the absence of things which directly correspond to mathematical concepts, he tells us that it suffices that figures exist which differ from true straight lines and circles to such a small extent that the error is smaller than any given error—"which is sufficient in order to demonstrate certainty as well as usage"[77].

10   Nature, Mathematics and Infinity

There are abundant remarks of this kind, reflecting Leibniz's conviction that nature can be shown to approach certain rational norms readily found in mathematics. The architectonics of the metaphysical model he developed around the concept of the monad in effect provides the foundation for the successful application of mathematics in modern scientific explanation, while conversely the success of the mixed sciences is interpreted as expressing not only divine benevolence but also a deep-rooted economy of the system of nature. Leibniz is clear, it seems to me, although I by no means want to ignore inconsistencies, even contradictions in his thought, that the monadological theory of centres of force founds a mechanistic model of nature which can be grasped by certain mathematical tools. How this application ties up with his metaphysics is not always clear. It is perhaps conceivable that Leibniz thought a description in general terms would largely suffice. I do not know. But be that as it may, it is precisely this combination of foundational metaphysics and already proven scientific achievements which enabled Leibniz to utter the expectation that by means of his calculus natural phenomena will in future be more adequately grasped than had hitherto been the case:

---

[75] Leibniz 1993, p. 39, (*De quadratura arithmetica circuli*, prop. VII, def.): "Et proinde si quis assertiones nostras neget facile convinci possit ostendendo errorem quovis assignabili esse minorem, adeoque nullum." See also Beeley 1999, p. 141.

[76] Leibniz 1993, p. 33 (*De quadratura arithmetica circuli*, prop. VI, schol.): "[…] cum nihil sit magis alienum ab ingenio meo quam scrupulosae quorundam minutiae in quibus plus ostentationis quam fructus, nam et tempus quibusdam velut caeremoniis consumunt."

[77] Leibniz, *De organo sive arte magna cogitandi*, A VI, 4, 159: "Quid autem de tribus his continuis sentiendum sit videtur pendere ex consideratione perfectionis divinae. Sed Geometria ad haec assurgere necesse non habet. Nam etiamsi non darentur in natura nec dari possent rectae ac circuli, sufficiet tamen dari posse figuras, quae a rectis et circularibus tam parum absint, ut error sit minor quolibet dato. Quod satis est ad certitudinem demonstrationis pariter et usus."

For this reason it is now not surprising that certain problems on receipt of my calculus have found solutions which earlier could scarcely have been hoped for and which especially concern the transition from geometry to nature. For traditional geometry is of little use as soon as the question of the infinite is involved, which is suitably involved in many operations in nature and whereby the Creator finds better expression.[78]

Leibniz is in fact more specific on how his calculus might be applied to solving physico-mathematical problems than here in his letter to John Wallis of 28 May 1697. A few months earlier in the same year he had outlined to Claudio Filippo Grimaldi in remarkably clear terms how he conceived his infinitesimal calculus would find application in nature, drawing among others on discussions concerning the distinction between primitive and derivative force in his *Specimen dynamicum* of 1695.

Now I have worked out a new method of expressing in a calculus infinitely small progressions of motions and the rudiments of these progressions which are infinitely infinitely small. Since motion, because it of course takes place in time is like a regular line, impetus as the momentary beginning of motion must be like an infinitely small or infinitesimal line. Thereupon conatus (for example gravity or the force receding from the centre), since it already constitutes impetus through infinite repetitions, will be an infinitely infinitely small quantity. When these notions of the infinite are transferred into a geometrical calculus the solutions to physico-mathematical problems which previously did not appear to be in our power but which now appear easier through infinite calculus are discovered, because nature everywhere involves something of the infinite, as it displays vestiges of the immeasurable author. Galileo investigated in vain the catenary line, which is formed when the most flexible chain consisting of the smallest links is suspended from two fixed points.[79]

---

[78] Leibniz to John Wallis, 28 May/[7 June] 1697, A III, 7, 430: "Unde jam mirum non est, problemata quaedam post receptum calculum meum soluta haberi, quae antea vix sperabantur; ea praesertim quae ad transitum pertinent a Geometria ad Naturam; quoniam scilicet vulgaris Geometria minus sufficit, quoties infiniti involvitur consideratio, quam plerisque Naturae operationibus inesse consentaneum est, quo melius referat Autorem suum." See also Leibniz, *De arte characteristica inventoriaque analytica combinatoriave in mathesi universali*, A VI, 4, 331; Beeley 2004, 24–41; Beeley 2009, pp. 38–39.

[79] Leibniz to Claudio Filippo Grimaldi, mid-January/mid-February 1697, A I, 13, 522: "Nam novam methodum excogitavi in calculo exprimendi motuum progressus infinite parvos et ipsa progressum rudimenta quae sunt infinities infinite parva. Cum enim motus, quippe qui in tempore fit, sit ut linea ordinaria, opportet ut impetus, tanquam initium motus instantaneum, sit ut linea infinite parva seu infinitesima. At conatus (exempli gratia, gravitas aut vis recedendi a centro) cum infinitis demum repetitionibus constituat impetum, erit quantitas infinities infinite parva. Has notiones infiniti ubi traduxi in calculum Geometricum inventae sunt solutiones problematum physico-mathematicorum, quae hactenus in potestate non esse videbantur, nunc autem paruere facilius calculo infinitesimali, quia natura ubique aliquid infiniti involvit, ut vestigia ostendat autoris immensi. Galilaeus frustra quaesierat lineam Catenariam, quam scilicet catena summe flexilis et annulis quam minimis constans, ex duobus datis punctis suspensa format." See also Leibniz to Schmidt, 3/13 August 1694, A I, 10, 499–500; Leibniz to Masson, n.d., GP VI, 629; Leibniz to Gabriel Wagner, [20/30 October 1696], A II, 3, 229: "wie ich […] mit meinem Calculo infinitesimali der Differenzen undt Summen die sach dahin gebracht, daß man in physico-mathematicis übermeistern kan, was man vor diesen anzutasten nicht einmahl sich erkühnen dürffen."; Leibniz, *Remarques de Mr. leibniz sur l'art. V. des Nouvelles de la République des Lettres du mois de février 1706*, GM V, 391; Leibniz, *De arte characteristica inventoriaque analytica combinatoriave in mathesi universali*, A VI, 4, 331; Leibniz, *Elementa nova matheseos universalis*, A VI, 4, 522.

Leibniz often cites his 1691 solution to the problem of the catenary in this context, since Johann Bernoulli had publicly called upon him to provide a solution by means of his calculus[80]. As is well known, Leibniz was subsequently able to claim to have achieved precisely this, when in June 1691 he sent his solution to Rudolph Christian von Bodenhausen for publication in Italy[81]. From a philosophical point of view Leibniz sees his infinitesimal calculus as lending itself not just to transcendental curves like the catenary or the Archimedean spiral, but also and in particular to grasping changes and processes in nature which as he points out always occur through an infinity of degrees and never through a discontinuous leap[82]. Moreover, he conceives such infinity of degrees explicitly in terms of inassignable or infinitely small quantities, whose infinite multitude must needs constitute finite, measurable quantities[83]. In this way, the architectonics of his metaphysical system and his mathematical practice again merge favourably, facilitating the passage from geometry to physics which he had expressed almost prophetically in his river metaphor of 1669. At the same time, almost incidentally, his law of continuity receives a more profound physico-mathematical foundation than it had at its first publication in 1687.

Conclusion

While Leibniz in his mathematical practice generally sought to steer clear of metaphysical disputes and instead placed emphasis on effectiveness and reliability of procedures, particularly when working with the infinite, he ultimately produced a philosophical system which in remarkable fashion was able to account for the successes of mathematical science in contributing to our understanding of nature and

---

[80] See Bernoulli 1690, 219.

[81] Leibniz to Bodenhausen, 12/22 June 1691, A III, 5, 118: "andere aber die solche [sc. Analysin novam] verachten und vor ein giocolino halten, können ihr heil an diesem problemate versuchen. Wiewohl nunmehr post exhibitam solutionem nichts leichter vor einen der den calculum verstehet, als rationem finden; aber ipsam solutionem zu finden soll einer wohl bleiben laßen, der nicht meinen oder einen aequivalenten Calculum hat."

[82] Leibniz to Thévenot, 24 August/[3 September] 1691, A I, 7, 355: "Comme c'est proprement l'Analyse des infinis que j'establis par ce nouveau calcul, et que la nature passe tousjours par une infinité degrés, et jamais per saltum, c'est par ce calcul, qu'on peut attribuer quelque chose à rompre la barriere, qu'il y a entre la physique et la Geometrie."

[83] Leibniz for Bodenhausen, *Änderungsvorschläge für Nova methodus als Anhang zu Dynamica*, A III, 4, 488: "Calculus incrementalis vel differentialis locum habet ubicunque mutationi locus est. Omnis autem mutatio in natura continua est, ac per gradus fit non per saltum, adeoque transitus fiunt per quantitates inassignabiles seu indefinite parvas, quarum infinita multitudine opus est ad quantitates communes constituendas."; Leibniz to Fardella, [3]/13 September 1696, A II, 3, 193: "Nam magnum inprimis usum habet calculus ille in transferenda Mathesi ad naturam, quia de infinito ratiocinari docet, omnia autem in natura habent characterem infiniti autoris."; Leibniz to Chauvin, 27 April/[7 May] 1697, A I, 14, 155: "La nouvelle science sert pour faciliter le passage de la Geometrie à la Physique, par ce que la consideration des effects de la nature enveloppe ordinairement l'infini pour exprimer le charactere de son auteur."

over and above this to the increase in human wellbeing. In a very true sense his philosophical deliberations on mathematics were deliberations pertaining to life[84]. As I have pointed out in the course of this paper, Leibniz incorporated infinite structures architectonically into his model of nature from the *Hypothesis physica nova* of 1671 onward. But he could scarcely have envisaged the degree to which the statement contained in his river metaphor would come to be theoretically verified over the years. His initial claims that mathematics finds application in nature, because the deviation is smaller than any quantity that can be given and because any resulting error must therefore be negligible are not wholly convincing until the concepts of negligible error and the infinitely small become embellished through the conceptual framework provided by infinitesimal calculus. It is almost as if Leibniz foresaw the very possibilities which his future work on infinitely small quantities would enable him and later mathematicians to achieve. In a very profound sense Leibniz's metaphysics was mathematical. Without doubt, concepts he devised in his *Theoria motus abstracti* stood him in good stead when he was developing his infinitesimal calculus. But perhaps, after all is said and done, this should not be so surprising to those of us who study his work. In his mathematization of nature Leibniz was quite simply exceptional, not just in the context of the early enlightenment. "It is a strange thing," he wrote to Duchess Elisabeth Charlotte of Orléans on 28 October 1696, "that one can calculate with the infinite just as one can with counters, and that nonetheless our philosophers and mathematicians have so little recognized how much the infinite is mixed up in everything"[85]. Leibniz, of course, did not count himself among them.

References

Arthur, R. 2003. The enigma of Leibniz's atomism. In: *Oxford studies in early modern philosophy,* eds. D. Garber and S. Nadler, vol. 1, 183–227. Oxford: Oxford University Press.

Arthur, R. 2009. Actual infinitesimals in Leibniz's early thought. In: *The philosophy of the young Leibniz,* eds. M. Kulstad, M. Lærke, and D. Snyder, 11–28. Stuttgart: Franz Steiner Verlag (Studia Leibnitiana Sonderheft 35).

Beeley, P. 1995. Les sens dissimulants. Phénomènes et réalité dans l'Hypothesis physica nova. In: *La Notion de Nature chez Leibniz,* ed. M. de Gaudemar, 17–30. Stuttgart: Franz Steiner Verlag (Studia Leibnitiana Sonderheft 24).

Beeley, P. 1996. *Kontinuität und Mechanismus. Zur Philosophie des jungen Leibniz in ihrem ideengeschichtlichen Kontext* . Stuttgart: Franz Steiner. (Studia Leibnitiana Supplementa vol. 30).

---

[84] Leibniz, *De numeris characteristicis ad linguam universalem constituendam*, A VI, 4, 268: "Hoc inprimis fit in deliberationibus ad vitam pertinentibus, ubi aliquid statuendum sane est, sed commoda atque incommoda (quae saepe utrinque multa sunt) velut in bilance examinare paucis datum est."

[85] Leibniz to Electress Sophie for Duchess Elisabeth Charlotte of Orléans, 28 October/[7 November] 1696, A I, 13, 85: "Et c'est une chose estrange, qu'on peut calculer avec l'infini comme avec des jettons, et que cependant nos Philosophes et Mathematiciens ont si peu reconnu combien l'infini est mêlé en tout."

Beeley, P. 1999. Mathematics and nature in Leibniz's early philosophy. In: *The young Leibniz and his philosophy* (1646–76), ed. S. Brown, 123–45. Dordrecht: Kluwer.

Beeley, P. 2004. In inquirendo sunt gradus—Die Grenzen der Wissenschaft und wissenschaftliche Grenzen in der Leibnizschen Philosophie. *Studia Leibnitiana* 36:22–41.

Beeley, P. 2009. Approaching infinity: Philosophical consequences of Leibniz's mathematical investigations in Paris and thereafter. In: *The Philosophy of the Young Leibniz,* eds. M. Kulstad, M. Lærke, and D. Snyder, 29–47. Stuttgart: Franz Steiner Verlag (Studia Leibnitiana Sonderheft 35).

Bernoulli, J. 1690. Analysis problematis antehac propositi. *Acta eruditorum*. May: 217–219.

Duchesneau, F. 1994. Leibniz on the principle of continuity. *Revue Internationale de Philosophie* 48:141–160.

Garber, D. 1995: Leibniz: Physics and philosophy. In: *The Cambridge companion to Leibniz,* ed. N. Jolley, 270–352. Cambridge: Cambridge University Press.

Garber, D. 2008. Dead force, infinitesimals, and the mathematicization of nature. In: *Infinitesimal differences. Controversies between Leibniz and his contemporaries,* eds. U. Goldenbaum and D. Jesseph, 281–306. Berlin: De Gruyter.

Gerhardt, C. I., ed. 1849–63. *Leibnizens Mathematische Schriften*. 7 vols. Berlin: A. Asher and Halle: H. W. Schmidt [cited as **GM**].

Gerhardt, C. I., ed. 1875–90. *Die Philosophischen Schriften von Gottfried Wilhelm Leibniz*. 7 vols. Berlin: Weidman [cited as **GP**].

Goldenbaum, U. 2008. Indivisibilia vera—How Leibniz came to love mathematics. In: *Infinitesimal differences. Controversies between Leibniz and his contemporaries,* eds. U. Goldenbaum and D. Jesseph, 53–94. Berlin: De Gruyter.

Hofmann, J. E. 1974. *Leibniz in Paris 1672–1676. His growth to mathematical maturity*. Cambridge: Cambridge University Press.

Huygens, C. 1669. A Summary Account of the Laws of Motion. *Philosophical Transactions* No. 46. 12 April: 927–928.

Knobloch, E. 1999. Im freiesten Streifzug des Geistes (liberrimo mentis discursu): Zu den Zielen und Methoden Leibnizscher Mathematik. In: *Wissenschaft und Weltgestaltung,* eds. K. Novak and H. Poser, 211–29. Hildesheim: Olms Verlag.

Knobloch, E. 2008. Generality and infinitely small quantities in Leibniz's mathematics—The case of his arithmetical quadrature of conic sections and related curves. In: *Infinitesimal differences. Controversies between Leibniz and his* contemporaries, eds. E. Goldenbaum and D. Jesseph, 171–83. Berlin: De Gruyter.

Leibniz, Gottfried Wilhelm. 1923. *Sämtliche Schriften und Briefe,* ed. Prussian Academy of Sciences (and successors); now: Berlin-Brandenburg Academy of Sciences and the Academy of Sciences in Göttingen. 8 series, Darmstadt (subsequently: Leipzig); now: Berlin: Otto Reichl (and successors); now: Akademie Verlag. [cited as **A**].

Leibniz, G. W. 1993. In *De quadratura arithmetica circuli ellipseos et hyperbolae cujus corollarium est trigonometria sine tabulis,* ed. E. Knobloch. Göttingen: Vandenhoeck & Ruprecht.

Mancosu, P., and E. Vailati. 1991. Torricelli's infinitely long solid and its philosophical reception in the seventeenth century. *Isis* 82:50–70.

Mancosu, P. 1996. *Philosophy of mathematics and mathematical practice in the seventeenth century*. Oxford: Oxford University Press.

Rutherford, D. 1995. *Leibniz and the rational order of nature*. Cambridge: Cambridge University Press.

The Difficulty of Being Simple: On Some Interactions Between Mathematics and Philosophy in Leibniz's Analysis of Notions

David Rabouin

1 Introduction

In a famous letter from November 1678, presumably written for Princess Elisabeth and dedicated to a critique of Descartes's philosophy, Leibniz states that he first cherished mathematics only because it was a way of accessing a more general *ars inveniendi*[1]. This is the path, he claims, on which he quickly realized the great limitation of Descartes's "method"[2]. One can easily acknowledge this limitation, Leibniz pursues, by considering how the Cartesian method supported a drastic restriction of the realm of mathematics—as opposed to the various projects of "new analyses" elaborated by Leibniz during his stay in Paris (such as: *analysis situs*, infinitesimal analysis, combinatorial analysis or *ars combinatoria*). The connection between mathematics and philosophy is made even tighter in the following passage of the letter from 1678 where, before turning to Descartes's demonstration of God's existence, Leibniz develops his objection by presenting now metaphysics itself as being not so different from the general *ars inveniendi* or "true Logic"[3].

However this famous text, supporting a harmonious connection between mathematics, logic and metaphysics holds some mysteries for anyone who tries to make this connection more explicit. What is the precise link between the limitation of the realm of mathematics and the limitation of the realm of metaphysics? What kind

---

[1] Unless otherwise stated, all the translations are mine.

[2] "Mais pour moy je ne cherissois les Mathematiques, que par ce que j'y trouvois les traces de *l'art d'inventer en général*, et il me semble que je découvris à la fin que Monsieur des Cartes luy même n'avoit pas encor penetré le mystère de cette grande science" (A II, 1, 662).

[3] "Car j'ay reconnu que la Metaphysique n'est gueres differente de la vraye Logique, c'est à dire de l'art d'inventer en general" (A II, 1, 662).

D. Rabouin (✉)
Laboratoire SPHERE, UMR 7219, CNRS—Université Paris Diderot, Paris, France
e-mail: davidrabouin2@gmail.com

© Springer Netherlands 2015
N. B. Goethe et al. (eds.), *G.W. Leibniz, Interrelations between Mathematics and Philosophy*, Archimedes 41, DOI 10.1007/978-94-017-9664-4_3

49

of extension is provided by the new forms of mathematical analysis to a "general" *ars inveniendi* or "true Logic"? How is this mathematical analysis supposed to be useful outside of mathematics? These are not easy questions to answer, especially as long as one keeps in mind that Leibniz had not yet developed at that time the kind of dynamical models which could support a direct—but perhaps misleading—analogy between mathematics and ontology (a certain "divine metaphysical mathematics" as he would later put it: *metaphysica quaedam mathesis*[4]).

To answer these questions, a particular attention must be given to the very specific mathematical examples mentioned in the 1678 letter. Indeed these examples are not taken from the "new analyses". They are notably impossibility results: the quadrature of the circle, the fastest speed, the "largest circle" and the "number of all units"[5]. My aim in this paper is to shed some light on this particular setting, put forward by the philosopher himself when mentioning the close connection between mathematics, logic and philosophy. In particular, I would like to bring certain mathematical documents dating from the stay in Paris into the picture and show their importance in the reflection on logical analysis (in the sense of "analysis of notions").

One consequence of the attention paid to these examples is, more generally, to bring new insights to what is sometimes presented as the core of Leibniz's philosophical program. As is well known, Leibniz once had the dream of reducing all human thoughts to a kind of computation based on an "alphabet"of simple notions (*quoddam Alphabetum cogitationum humanarum*)[6]. Next to the dynamical model, this is often presented as an essential feature of the way in which mathematics and philosophy are related in his work[7]. Indeed it does seem to command a certain similarity between human thoughts and mathematical "combinations" (whatever these "combinations" are). Symmetrically, it leads to a certain program of rethinking mathematics by considering the centrality of *ars combinatoria*. One famous goal

---

[4] To De Volder, September 6th 1700, GP II, 213.

[5] "Il faut avouer aussi que la preuve de Mons. des Cartes qu'il apporte à fin d'establir l'idée de Dieu est imparfaite. Comment dira-il pourroit on parler de Dieu sans y penser, et pourroit on penser à Dieu sans en avoir l'idée. Ouy sans doute, on pense quelques fois à des choses impossibles, et mêmes on en fait des demonstrations. Par exemple Mons. des Cartes tient que la quadrature du cercle est impossible, et on ne laisse pas d'y penser, et de tirer des consequences de ce qui arriveroit si elle estoit donnée. Le mouvement de la derniere vistesse est impossible dans quelque corps que ce soit (…). De même le plus grand de tous les Cercles, est une chose impossible, et le nombre de toutes les unités possibles ne l'est pas moins: il y en a démonstration." (A II, 664).

[6] "I (…) arrived at this remarkable thought, namely that a kind of alphabet of human thoughts can be worked out and that everything can be discovered and judged by a comparison of the letters of this alphabet and an analysis of the words made from them. This discovery gave me great joy though it was childish of course, for I had not grasped the true importance of the matter" (*De numeris characteristicis ad linguam universalem constituendam* (1679), A VI, 4, 265; transl. Loemker (1989b), 222).

[7] This program was of prime importance in the picture of Leibniz's thought drawn by Couturat (1901)—especially chap. VI: "La Science Générale", 180–184, where Couturat summarizes his interpretation by commenting on a fragment on which I shall also put some emphasis: *De la Sagesse* [(1676); A VI, 3, 671]. On the relation between this program and that of a *Characteristica Universalis*, its origins and sources, see also Rossi (2006) and Pombo (1987).

associated with this program was that of a *Characteristica Universalis*: a way of representing all human knowledge on the model of a combination of basic symbols representing primitive notions[8]. What is less well known, however, is that Leibniz expressed very early on serious doubts about the feasibility of such a program and that these doubts emerged from his mathematical practice. Moreover, these doubts are closely related to the arguments provided in the Letter to Elisabeth and attached to the problems raised by the demonstrations of impossibility. My goal, in this paper, is to shed some light on this specific interaction between mathematics and philosophy and to indicate its importance in Leibniz's philosophical evolution.

1.1   "Leibniz's Program": Toward an Alphabet of Human Thoughts

According to Leibniz's own account, his grand program emerged when he was a young boy from a very straightforward dissatisfaction:

> As a boy I learned logic, and having already developed the habit of digging more deeply into the reasons for what I was taught, I raised the following question with my teachers. Seeing that there are categories for the simple terms by which concepts are ordered, why should there not also be categories for complex terms, by which truths may be ordered? (*De synthesi et analysi universali seu Arte inveniendi et judicandi*; A VI, 4, 538; transl. Loemker (1989b, 229))

From here emerged Leibniz's suspicion that traditional categories were not well constituted and that there should be a process of constructing *praedicamenta* from simple notions in a more accurate way. This process could then be reproduced on each level and mimic the production of words and sentences from an alphabet (which Leibniz saw as a kind of "combination"). This is the famous theme of the "alphabet of human thoughts":

> It seemed to me (...) that this could be achieved universally if we first had the true categories for simple terms and if, to obtain these, we set something new in the nature of an alphabet of human thoughts, or a catalogue of the highest genera or of those we assume to be the highest, such as *a, b, c, d, e, f,* out of whose combination inferior concepts may be formed (*ibid.*).

This account, written in the 1680s, is not merely a historical reconstruction by Leibniz. On the contrary, it refers directly to the *De Arte Combinatoria* (1666), and particularly to the sections concerned with the application (*usus*) of problems dealt with in the first part of the treatise. The general program is presented there as a renewal of the "analytical part" of Logic (*partem logices Analyticam*) and one of the main points of reference is the composition of propositions through basic "names"

---

[8] "We can make the analysis of thoughts perceptible and conduct it as if by some mechanical guide, since the analysis of characters is something that is somewhat perceptible. Indeed *Analysis of characters* occurs when we substitute certain characters for others, which are equivalent to the former in their use." (*Analysis Linguarum* (September1678); A VI, 4, 102).

proposed by Hobbes at the beginning of the *De Corpore* (GP IV, 46). In the tenth application, after a long discussion of Lull's art, Leibniz makes a more explicit reference to the Hobbesian model as a kind of computation; he then proposes a numerical model, based on the parallel between the decomposition of a term and the decomposition of a number into its factors (GP IV, 64–66). This analogy is also transcribed through an alphabetical model and the project of a *scriptura universalis* (GP IV, 72).

Even if these connections are still based on mere analogies, they nonetheless give the main direction of a program which Leibniz would describe, in a letter written to Duke Johann Friedrich in 1671, as "to do in philosophy what Descartes and others have done in arithmetic and geometry by means of algebra and analysis" (A II, 1, 160). Although "analysis" has a very broad meaning in Leibniz's work and although it is not clear that one can easily seize a unity behind this term, this program provides us with a first coherent approach to a sound understanding. It should be noted that the term "analysis" is used in the *De Arte Combinatoria* not only to designate the "analytical part" of logic based on "analysis of notions", but also symbolical algebra as it was used to solve geometrical problems after Descartes[9]. From what we have seen above, the analogy between mathematical and logical "analysis" is first grounded on the basic operation of a decomposition into (an alphabet of) simple entities and (re)composition (of truths) on the basis of rules of combination or computation.

It is very tempting to see this first set of features as an ideal which Leibniz pursued throughout the whole of his philosophical work. Reconsidering the *De Arte Combinatoria* in 1691, after it had been republished without his consent, Leibniz still acknowledged its importance in having spread the first seeds of the true *ars inveniendi*—the same topic as in the letter to Elisabeth. He clearly related those early efforts to the general project of an "analysis of thoughts" by means of a reduction of complex propositions to an alphabet of primitive notions[10]. Moreover, Leibniz not only continued exploring different ways of constituting a catalog of basic notions up to the end of his life—be it in projects of a general Encyclopedia or in *initia* of different sciences, in particular Geometry—, but he also elaborated a theory of knowledge, which gave a very important role to this "analysis of thoughts".

As is well known, this theory was first publicly presented in the *Meditationes de cognitione, veritate et ideis*, although it had already emerged during the stay in Paris[11]. It subsisted, at least partially, until the time of the *Nouveaux Essais*. Its basic structure was given by a series of criteria meant to guide the "decomposition" of notions and to provide a precise meaning for the terms used by the Cartesians

---

[9] "This is the origin of the ingenious specious analysis which Descartes was the first to work out, and which *Frans van Schooten* and *Erasmus Bartholinus* later organized into precepts; the latter in what he calls the *elements of a universal mathematics. Analysis* is thus the science of ratios and proportions, or of unknown quantity, while *arithmetic* is the science of known quantity, or numbers" (GP IV, 35; transl. Sasaki (2003, 298)).

[10] *Novas complures meditationes non poenitendas, quibus semina artis inveniendi sparguntur, (…) atque inter caeteras palmariam illam de Analysi cogitationum humanarum in Alphabetum quasi quoddam notionum primitivum* (A VI, 2, 549).

[11] See Picon (2003).

when they claimed that one should rely on "clear and distinct ideas" and try to reduce complex judgment to them[12]. According to Leibniz, a notion is *clear* when we can recognize it without necessarily being able to decompose it into its distinctive characteristics; it is *distinct* when we can decompose it into its distinctive characteristics (i.e. what suffices to distinguish it from other notions) or when it is its own characteristic (i.e. when it is already a primitive notion, with no prerequisite). In the case of a complex notion, to have a distinct knowledge of that notion does not mean, however, that we actually possess a distinct knowledge of the characteristics entering into it. This kind of knowledge, although distinct, can therefore still be *symbolic* or *blind*, but "if all which enters into a distinct knowledge is in its turn distinctly known, i.e. if the analysis can be conducted through to an end, then this knowledge is *adequate*"[13].

As is obvious from this description, *adequate* knowledge is not possible without concomitant knowledge of the fact that the analysis is *complete*. This seems to imply that we have at the same time access to *primitive notions*, i.e. notions which are their own characteristic. In this picture, the "alphabet of human thoughts" is not just a program: its realization, or at least partial realization, seems a necessary *condition* for *any* adequate knowledge whatsoever. This would explain why Leibniz spent a lot of time and energy trying to isolate what he calls the *initia* of different sciences, especially in mathematics, where it was supposed to be easier to realize.

This "analytical" model is also present in the way Leibniz pictures not only notions, but propositions and reasoning. This is the ground for the famous argument according to which many truths which could appear to be "self evident" are in fact demonstrable. One should always try, Leibniz argues, to analyze truths further until one reaches "identicals" (for example: tautologies of the form *A is A*). Thus a famous passage of the *Nouveaux Essais* shows that propositions like "2+2=4" are not immediate and can be proved by means of definitions and axioms[14]. More generally, the very definition of a necessary truth or "truth of reason" is tied to its "analytical" nature: "When a truth is necessary, its reason can be found by analysis, resolving it into simpler ideas and simpler truths until we reach the primitives" (*Monadology* § 33; GP IV, 612). Patching these different pieces together, one could then easily come to the conclusion that Leibniz never gave up a certain "analytical"

---

[12] More than targeting at Descartes himself, the model of "Cartesian" logical analysis aimed at by Leibniz here is the one provided by Arnauld and Nicole in their *Logique* (1662).

[13] *Meditationes de Cognitione, veritate et ideis* (A VI, 4, 587).

[14] "Two and two are four is not quite an immediate truth. Assume that four signifies three and one. Then we can demonstrate it, and here is how.

Definitions:
(1) *Two* is one and one
(2) *Three* is two and one
(3) *Four* is three and one
Axiom: If equals be substituted for equals, equal remains.
Demonstration:

2 and 2 is 2 and 1 and 1 (def. 1) $2+2$
2 and 1 and 1 is 3 and 1 (def. 2) $(2+1)+1$
3 and 1 is 4 (def. 3) $3+1$

Therefore (by the axiom) 2 and 2 is 4, which is what was to be demonstrated

model of "perfect knowledge" based on the possibility of decomposing complex judgments into simpler ones, until one reaches primitive notions[15].

While I do not want to deny that this picture was very important in Leibniz's thought and that it remained in the background of many of his projects, my aim in this paper is to bring into this picture some nuances. I want to show not so much that this kind of model is not pertinent, but that it is only one amongst *other* models of "perfect knowledge". In particular, one should not conflate the reduction to "identicals" as presented in the model of "analytical" proofs with the reduction to "simple notions" as characterizing accurate knowledge, although the two may possibly coincide (when the "identicals" under consideration happen to involve primitive notions)[16]. As I will argue in the following section, Leibniz saw very quickly how difficult, if not impossible, it was to reach a complete analysis of notions—even in the simple case of natural numbers. Reflecting on this issue, he came to realize that knowledge can be perfect *without* a complete analysis of the notions involved in it and he elaborated other models of logical analysis.

1.2 Doubts

My goal is not to enter into the details of what I have presented *cum grano salis* as "Leibniz's program", but to emphasize first that this program underwent *very early* a profound critique by its alleged main proponent. For example, in a text dedicated to the "Elements of thoughts" (*De Elementis cogitandi*), which was presumably written in the spring of 1676, Leibniz starts by claiming that "if it is true that there is a perfect demonstration, i.e. one which leaves nothing unproven, then it *necessarily* follows that some elements of thought must exist, since the demonstration will be perfect *only when everything is analysed*". This seems to be an orthodox description of what we have seen as his "grand program". However, he adds immediately a startling *caveat*: "*But I realize now that this is false*"! He then explains this quite radical change of view in the following way: "a demonstration is perfect as soon as one can reach identical [propositions], which can happen *even though everything is not analysed*. For even notions which are not absolutely simple (like the parabola, or the ternary) can be stated from one another"[17]. In the terminology of the *Meditationes*, it is possible to have *true* knowledge which is not *adequate*

---

[15] As is well known, Leibniz never uses the expression "analytical proposition" which would come to have such an importance in later philosophy. Nonetheless, he talks of "analytical truths" (*analyticae veritates*) in order to designate truths which can be analyzed into simple terms and therefore expressed by natural numbers (A VI, 4, 715).

[16] The role of the reduction to "identicals" in Leibniz's mathematical practice is the subject of another paper of mine, which could be considered as a continuation of the present one cf. Rabouin (2013).

[17] A VI, 3, 504; my emphasis. Note that Leibniz explicitly mentions a case in which the notions are *not* analyzed in contrast to a case in which they would be *assumed* to be already analyzed, i.e. free of any potential contradiction.

and this possibility is opened by the very existence of *symbolic* thinking. This is of importance, since according to the *Meditationes*, we make use of symbolic thinking "almost everywhere": "in fact, if the concept is highly composed, we *cannot* think of all the notions entering it at the same time" (A VI, 4, 588; my emphasis).

In a paper written during the same period as *De Elementis cogitandi* and dedicated to investigating the "first propositions and first terms", Leibniz sets out another important warning:

> We cannot easily recognise indefinable primary terms for what they are. They are like prime numbers, which we have hitherto been able to identify only by trying to divide them. *Similarly, irresoluble terms could be recognised properly only negatively and provisionally.* For I know one criterion by which one can recognise resolubility. This is as follows: when we come across a proposition which looks necessary, but has not been demonstrated, it infallibly follows that this proposition contains a definable term (provided that it *is* necessary). So we must try to give this demonstration, which cannot be done without finding the definition in question. By this method, letting no axiom go without proof (except definitions and identicals), we shall arrive at the resolution of terms, and the ultimately simple ideas. ("Sur les premières propositions et les premiers termes" (1676); A VI, 3, 436; my emphasis)

Interestingly enough, he then draws a parallel with the empirical sciences:

> You will say that this could go on to infinity, and that new propositions could always be proved, which would oblige us to look for new resolutions. I do not believe so. But if it were the case, it would not matter, since by this method *we would not have failed to have perfectly demonstrated all our theorems,* and the resolutions which we would have performed would suffice us for an infinity of valid practical inferences. Just as in natural science we should not abandon experimental research because of its potential infinity, since we can already make perfectly good use of the results we have so far obtained. (*ibid.*; my emphasis)

These texts seem to indicate that Leibniz gave up very early the dream of establishing *first* a good connection between notions and judgment, *before* pursuing the reform of the *ars inveniendi*. This is quite explicit in another fragment from the same period in which Leibniz carefully distinguishes between what he calls "analyse des choses" and "analyse des vérités":

> It is very difficult to achieve the analysis of things, but it is not so difficult to achieve the analysis of truths needed. Since analysis of truth is achieved as soon as one has found a demonstration: and *it is not always necessary to achieve the analysis of the subject or of the predicate to find the demonstration of a proposition.* Most of the time, the beginning of the analysis of a thing is enough to reach the analysis or *perfect* knowledge of the truth that we get of this thing[18].

One could consider this amendment not to be a substantial one: it could just state that the analysis of notions and the analysis of truths can be conducted separately, *before* they are merged in the final state of the research. But the *Meditationes* makes

---

[18] "Il est très difficile de venir à bout de l'analyse des choses, mais il n'est pas si difficile d'achever l'analyse des vérités dont on a besoin. Parce que l'analyse d'une vérité est achevée quand on en a trouvé la démonstration: *et il n'est pas toujours nécessaire d'achever l'analyse du sujet ou du prédicat pour trouver la démonstration de la proposition.* Le plus souvent le commencement de l'analyse de la chose suffit à l'analyse ou connaissance *parfaite* de la vérité qu'on connaît de la chose" (*De la Sagesse* (1676); A VI, 3, 671; my emphasis).

it clear, by mentioning the case of notions entailing a hidden contradiction, that a provisional separation was highly problematic. I will come back to this situation later since it is precisely the context in which mathematical proofs of impossibility reveal their importance in the analysis of notions. But let me first emphasize the consequence of this fact: if true knowledge can be symbolic, it should not be forgotten that *prima facie* instances of symbolic knowledge can turn out to be *false*, i.e. that we can manipulate clear and distinct notions which, after analysis, reveal a contradiction. This leads to a rather dramatic conclusion: there is no guarantee that analysis of notions and analysis of truth, when conducted separately, can finally merge into harmonious universal knowledge. Analysis of truth, despite its name, does not give access to the true "content" of the propositions involved. It is always possible that a contradiction lies hidden in the combination of notions involved in the reasoning. This is precisely one of the major objections that Leibniz raised against Descartes' demonstration of God's existence.

Leibniz's position might have been recognized more clearly, had there not been the unfortunate tendency to cut the quotations at the crucial moment where Leibniz states explicitly the kind of *caveat* we encountered in the texts from 1675 and 1676. Let us return to the *Meditationes* where knowledge of numbers serves as the *only* example of adequate knowledge. This example provides an interesting connection to the theme of analytic truth presented in the *Nouveaux Essais* and, at the same time, gives the impression that we have here a very simple case of *perfect* analytical knowledge. But it is rare that Leibniz's precise declaration is given in its entirety: "if the analysis can be conducted through to an end, then the knowledge is *adequate*; whether man can give a perfect example of this *I do not know*, even if the knowledge of numbers certainly *comes very close to it*" (A VI, 4, 587; my emphasis).

The same should be said of the often quoted passage from "La vraie méthode" (1677) where Leibniz states his famous *calculemus* and describes what is often presented as the background to his general program, the possibility of reducing all human controversies to some computations:

> Whence it is manifest that if we could find characters or signs appropriate for expressing all our thoughts as definitely and as exactly as arithmetic expresses numbers or geometric analysis expresses lines, we could in all subjects in so far as they are amenable to reasoning accomplish what is done in Arithmetic and Geometry. For all inquiries which depend on reasoning would be performed by the transposition of characters and by a kind of calculus, which would immediately facilitate the discovery of beautiful results. For we should not have to break our heads as much as is necessary today, and yet we should be sure of accomplishing everything the given facts allow. Moreover, we should be able to convince the world what we should have found or concluded, since it would be easy to verify the calculation either by doing it over or by trying tests similar to that of casting out nines in arithmetic. And if someone would doubt my results, I should say to him: 'Let us calculate, Sir' and thus by taking to pen and ink, we should soon settle the question. ("La vraie méthode", 1677; A VI, 4, 3; transl. Wiener, 15)

This is a beautiful program indeed, but it is one which seems dramatically mutilated if we do not also read the following passage: "I still add: *in so far as the reasoning allows on the given facts*. For although certain experiments are always necessary to serve as a basis for reasoning, nevertheless, once these experiments are given, we should derive from them everything which anyone at all could possibly derive" (*ibid.*). The emphasis on the necessity of certain experiments as basis for knowledge is surprising in this context and modifies significantly the picture given when mentioning the first part of the text only.

All these passages lean in the same direction: Leibniz supported a much more nuanced and flexible conception of the organization of knowledge than that which could be inferred from the "program" sketched in Sect. I.1. In particular, he is very clear, at least after his arrival in Paris, that *perfect* knowledge—the term occurs regularly in our quotes—can be achieved *without* a complete analysis of notions. More than that, it is not even established that we may possess *one single example* of a truly complete analysis of notions: *exemplum perfectum nescio an homines dare possint*. My aim, in the next sections, is to underline the role played by mathematical practice in the development of these ideas. By so doing, I hope to shed light on the declaration to Elisabeth from 1678 about the connection between mathematics and the constitution of a general *ars inveniendi* or "true logic".

2 Behind the Doubts

In this section, I will put particular emphasis on two types of mathematical examples: first, we have the role played by the impossibility results mentioned to Elisabeth, especially the "number of all numbers" and the "quadrature of the circle"; second, we have the research undertaken in number theory. This latter example is mentioned in similar texts than the letter to Elisabeth such as the letter to Malebranche written in January 1679[19]. One common point to these examples is that they are taken neither from the new "transcendental geometry" nor from *ars combinatoria*, which are more often than not considered as the main, if not only, mathematical influences in Leibniz's philosophical reflections. By focusing on other practices, I hope to give not only a larger view of Leibniz's conceptions, but also a picture closer to that he presented in 1678 regarding the role of mathematics in the constitution of a general *ars inveniendi*.

2.1 The "Number of all Numbers"

It is well known that Leibniz had no real training in mathematics before he arrived in Paris and that his mathematical awakening was partly due to his encounter with Christiaan Huygens. It is Huygens who opened the eyes of the young German scholar to the modern mathematical literature and who talked to him about a problem he encountered when dealing with probabilities: how to prove that the sum of

---

[19] A II, 1, 677; see below note 46.

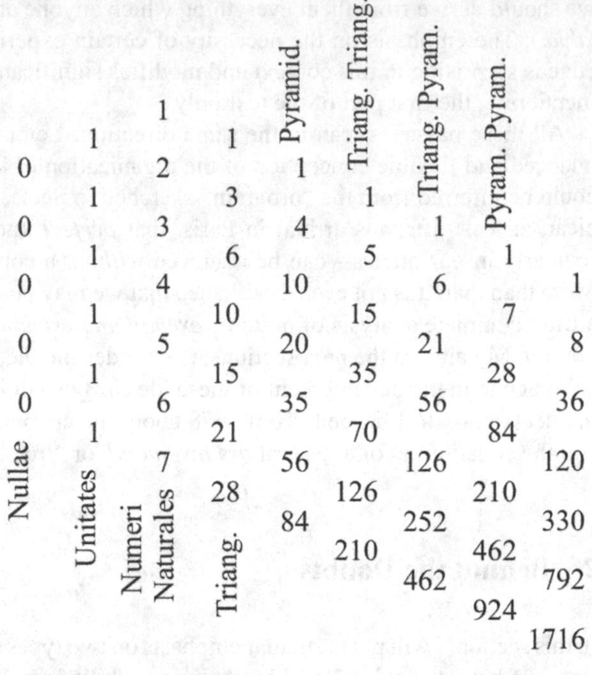

Fig. 1 Pascal's Triangle as transcribed by Leibniz in the *Accessio*. (A II, 1, 345–346)

Nullae	Unitates	Numeri Naturales	Triang.	Pyramid.	Triang. Triang.	Triang. Pyram.	Pyram. Pyram.	
		1						
	1		1					
0		2		1				
	1		3		1			
0		3		4		1		
	1		6		5		1	
0		4		10		6		1
	1		10		15		7	
0		5		20		21		8
	1		15		35		28	
0		6		35		56		36
	1		21		70		84	
		7		56		126		120
			28		126		210	
				84		252		330
					210		462	
						462		792
							924	
								1716

the inverse of the "triangular numbers", i.e. the third line in what is now called the "Pascal Triangle", is equal to two (Fig. 1).

Generalizing techniques regarding the summation of infinite series found in Grégoire de Saint Vincent, Leibniz was able not only to give an answer to Huygens' problem (which the Dutch mathematician already knew by other means), but also to calculate all of the sums in the "inverse" Pascal Triangle, which he would later coin the "Harmonic Triangle"[20]. Huygens was evidently impressed by these results, coming as they did from a near beginner in mathematics. For his part, Leibniz thought that results which impressed one of the greatest mathematicians of the time would suffice to insure his glorious entrance into the Parisian mathematical community[21]. The result of these early efforts was transcribed in a text written at the end of 1672 in the form of a letter to Jean Gallois, the editor of the *Journal des Sçavans* and entitled *Accessio ad arithmeticam infinitorum*[22]. It is in this text that one can find the Leibnizian proof of the impossibility of a "number of all numbers".

---

[20] On the methods used by Leibniz see Hofmann (1974, 15–20).

[21] This was of course before his first journey to London where he came to realize just how limited his knowledge of mathematics in general, and of series in particular, actually was. See S. Probst's paper in this volume.

[22] A II, 1, 342–356. Part of the story is told by Leibniz himself at the beginning of the text. In fact, the paper was not submitted because the journal ceased publication for two years after December 1672.

Fig. 2 The sums of the reciprocal figurate numbers. (A II, 1, 347)

Series Fractionum Progressionis Arithmeticae Replicatae

	semel	bis	ter	quater	quinquies	sexies	septies
Exponentes	1	2	3	4	5	6	7
	Unitatum	Naturalium	Triangularium	Pyramidalium	Triangulo-Triangularium	Triangulo-pyramidalium	Pyramido-pyramidalium
	$\frac{1}{1}$	$\frac{1}{1}$	$\frac{1}{1}$	$\frac{1}{1}$	$\frac{1}{1}$	$\frac{1}{1}$	$\frac{1}{1}$
	$\frac{1}{1}$	$\frac{1}{2}$	$\frac{1}{3}$	$\frac{1}{4}$	$\frac{1}{5}$	$\frac{1}{6}$	$\frac{1}{7}$
	$\frac{1}{1}$	$\frac{1}{3}$	$\frac{1}{6}$	$\frac{1}{10}$	$\frac{1}{15}$	$\frac{1}{21}$	$\frac{1}{28}$
	$\frac{1}{1}$	$\frac{1}{4}$	$\frac{1}{10}$	$\frac{1}{20}$	$\frac{1}{35}$	$\frac{1}{56}$	$\frac{1}{84}$
	$\frac{1}{1}$	$\frac{1}{5}$	$\frac{1}{15}$	$\frac{1}{35}$	$\frac{1}{70}$	$\frac{1}{126}$	$\frac{1}{210}$
	$\frac{1}{1}$	$\frac{1}{6}$	$\frac{1}{21}$	$\frac{1}{56}$	$\frac{1}{126}$	$\frac{1}{252}$	$\frac{1}{462}$
	$\frac{1}{1}$	$\frac{1}{7}$	$\frac{1}{28}$	$\frac{1}{84}$	$\frac{1}{210}$	$\frac{1}{462}$	$\frac{1}{924}$
Summae	$\frac{0}{0}$	$\frac{1}{0}$	$\frac{2}{1}$	$\frac{3}{2}$	$\frac{4}{3}$	$\frac{5}{4}$	$\frac{6}{5}$ etc.

The *Accessio* consists of a remarkable mixture of mathematical and philosophical reflections. The first part, after a general summary of what Leibniz thought to be the knowledge on infinite series at that time, presents his own results, namely the calculation of the sums for all the columns of the harmonic triangle (Fig. 2). He emphasizes the fact that he can then produce a general rule: the sum of all numbers in a column of the harmonic triangle is equal to the ratio of the "exponents" of the two preceding ones. The third column being of exponent 3, its sum will therefore be equal to $2/1$[23].

As can be seen in the above table, Leibniz's *regula universalis* led him to an audacious interpolation: the sum of the first column, which is the sum of all units, should be equal to the "exponent" of the preceding column over the preceding one, which gives $\frac{0}{0}$ (Leibniz does not consider that there could be a "-1"-column; he also interprets $\frac{0}{0}$ as giving the value 0). This is the starting point of the philosophical reflections developed in the second part of the paper. In it, Leibniz dwells upon

---

[23] As Pascal already put it, the rule of formation for the triangle does not hold that the first line be generated by unity. Leibniz therefore gives a more general result concerning any series of fractions formed with an arbitrary generator of a Pascal Triangle as numerator and a line of the triangle as denominator: "*Regula Universalis haec est: Summa seriei fractionum, quarum numerator est generator, nominatores sunt termini cujusdam progressionis Arithmeticae Replicatae est fractio seu ratio cujus numerator seu antecedens (…) est exponens seriei proxime praecedentis seu penultimae (data scilicet supposita ultima) nominator vero seu consequens est exponens seriei proxime praecedentis praecedentem, seu antepenultimate*" (A II, 1, 346).

the fact that he has produced a demonstration that the "sum of all units", i.e. the "maximal number" or "the number of all numbers" (since any natural number can be expressed as a certain finite sum of units) amounts to ... "nothing"[24]. It therefore represents an impossible notion (the extension of it being empty).

His main goal in the second part of the letter is then to refute Galileo's argument, based on the famous paradox which now bears his name[25]. The Pisan mathematician considered that infinity was comparable to unity and was therefore a "non-quanti-ty". His conclusion rested on the fact that in infinity the axiom *totum esse majus parte* fails. Leibniz's own proposal is that infinity is more properly comparable to zero and therefore the axiom "the whole is greater than the part" holds universally. In effect, the only exception which can be found amounts to "nothing". This belief is also grounded on the fact that one can produce a *demonstration* of this so-called axiom (which Leibniz undertakes to give at the close of the letter).

Leibniz's enthusiasm for this result is quite surprising for the modern reader, not only because it is by far the least convincing result of the paper, but also because at first glance it seems mathematically quite trivial. Indeed one could consider that the impossibility of the greatest number which would be the "sum of all units" is evident by the construction of natural numbers. If not, it would, in any case, be a simple consequence of Euclid's *Elements* IX.20 which proves that "the prime numbers are more than an assigned multitude of prime numbers"[26]. How could Leibniz be so proud to have produced a proof of the fact that there is no largest number? I would propose two possible and complementary replies:

- first, one could argue that Euclid's and Leibniz's proofs are fundamentally different in that the first proves that there is a potential infinity of prime numbers (and therefore of natural numbers) whereas the second proves that there cannot be an actual infinity of them[27].
- second, one could argue that Leibniz was impressed not so much by the mathematical but by the philosophical consequences of this result.

The first answer is largely confirmed by the fact that Leibniz, although he would rely on other demonstrations, would always mention this result in later texts in the context of the denial of the existence of an actual infinite in mathematics[28].

---

[24] "*Numerum istum infinitum sive Numerum maximum seu omnium Unitatum possibilium summam, quam et infinitissimum appellare possis, sive numerum omnium numerorum esse 0 seu Nihil*" (A II, 1, 352).

[25] This "paradox" is modeled on the fact that one can establish a one to one correspondence between natural numbers and their squares (or their cubes), where it seems that there are "more" of the former than the latter (not all natural numbers are squares). See S. Levey's paper in this volume.

[26] I thank Marco Panza for having pressed me on this issue when I first presented this study.

[27] As is well known, Euclid's proof states that whatever prime number is considered to be the largest, it is always possible to construct a larger one.

[28] See, for example, the letter of late August 1698 to Johann Bernoulli: "Many years ago I proved beyond any doubt that the number or multitude of all numbers implies a contradiction, if taken as a unitary whole. I think that the same is true of the largest number, and of the smallest number, or

The second answer emerges naturally from the last section of the letter, devoted to a lengthy discussion of the fact that one should not accept any self evident truth except definitions, being as they are mere postulations[29]. In particular, one should always try to find demonstrations of the axioms, following the example of the preceding discussion on the axiom *totum esse majus parte*.

In other places Leibniz would point to further philosophical consequences closely related to the result mentioned in the letter to Gallois. An important testimony is given in the letter to Oldenburg, dated 28 December 1675, in which he states:

> But we seem to think of many things (though confusedly) which nevertheless imply [*scil.* contradiction]; for example, the number of all numbers. Hence we should be strongly suspicious about notions like infinity, minimum and maximum, the most perfect, and even totality itself. Nor should we trust these notions until they have been measured by that criterion, which I seem to recognize, and by which truth is rendered stable, visible, and, so to speak, irresistible as if it were by a mechanical procedure.[30]

We find here the general strategy which is at the core of Leibniz's argument against Descartes in the letter to Elisabeth from 1678—and also at the center of the argumentation in the *Meditationes*: some notions like "the maximum of perfection" entering into the so-called "ontological proof" ought to be treated with caution since they may entail, after logical analysis, some hidden contradiction. When manipulating them, one should therefore always accompany the "blind" reasoning with a demonstration of their possibility (or impossibility). The fact that the *only* example given to Oldenburg is that of the "number of all numbers" confirms the crucial role played by this mathematical example in the development of Leibniz's philosophical ideas.

Let me develop this aspect in more detail and thereby consider for the sake of argument that Leibniz's proof in the *Accessio*, as he thought it, is valid. What does it indicate in terms of theory of knowledge? This is not obvious if we do not keep in mind the general setting of Leibniz's initial "program". The core of this program, as we have seen, is the following: if we succeed in analyzing our knowledge until we reach simple notions and if we are very cautious about combining them according to simple intuitive relationships, we should succeed in mastering the totality of all potential knowledge—which could be acquired by mere "combinations" or computations[31]. Now, what is the relationship between this general philosophical program

---

the lowest of all fractions. The same has to be said about these, as about the fastest motion and the such-like" (GM III, 535, transl. MacDonald Ross (1990, 129)). On the impossibility of an actual infinite number, see also *Essais de Théodicée* (GP VI, 90).

[29] *"exceptis scilicet ipsis definitionibus, quae ut toties in suis scriptis inculcat restaurator philosophiae Galilaeus, arbitrariae sunt, nec falsitatis, sed ineptiae obscuritatisque tantum arguendae"* (A II, 1, 351).

[30] *Multa videmur nobis cogitare (confuse scilicet) quae tamen implicant. Exempli causa: Numerus omnium numerorum. Unde valde suspecta esse debet nobis notio infiniti, et minimi et maximi, et perfectissimi, et ipsius omnitatis. Neque fidendum his notionibus, antequam ad illud criterion exigantur, quod mihi agnoscere videor, et quod velut Mechanica ratione, fixam et visibilem, et ut ita dicam irresistibilem reddit veritatem* (A II, 1, 393).

[31] Even if it is dubious that this conception was genuinely Cartesian, it is certainly very close to the kind of representation of knowledge presented in the *Regulae ad directionem ingenii*. In the

Fig. 3 A regular decahedron?

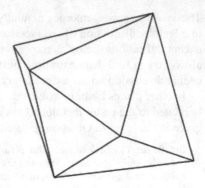

and the result put forward in the *Accessio*? It is, I think, quite clear: the *numerus maximus* as the "sum of all units" gives us a very simple counter-example of a notion which can be decomposed into apparently simple intuitive notions (unity) and intuitive relations (addition of units), but of which the synthesis is nonetheless *impossible*. This is precisely what Leibniz explains to Oldenburg: one should be very cautious with some notions that look perfectly meaningful, but which, after careful analysis, reveal themselves to be impossible.

One could object that the problem here comes from the recourse to infinity in the process of synthesis. Because of this recourse to infinity, it would be a typical case for which the "Cartesian" model of analysis would *not* claim to work. But Leibniz would have no difficulty in formulating other examples of impossible notions which do not involve recourse to infinity. This is the case of the famous regular decahedron mentioned in the *Nouveaux Essais sur l'entendement humain* (III, 3, § 15; GP V, 272). The situation is exactly the same: I can think of a regular decahedron, define it clearly and distinctly as a polyhedron composed of ten identical faces, break it down into apparently simple intuitive notions, namely segments forming ten triangles glued to one another in a diamond shape (such ten sided diamonds do exist in nature) (Fig. 3). More than that, I can produce true reasoning about it, such as calculating the relationship between the number of vertices, edges and faces—what we now would call its "Euler number". This is a very nice and simple example of the danger of "blind" or "symbolic" thinking. In this case, it entails, in effect, an impossible notion: there is no such thing as a regular decahedron

---

Regulae Descartes reduced the entirety of knowledge, at least such as is accessible to certitude, to two basic operations: *intuition* and *deduction* (these being considered as chains of directly evident inferences). Although Leibniz read the *Regulae* only at the beginning of 1676, this model was clearly presented by the *Logique de Port-Royal* (1662) as the paradigm of the new, that is "Cartesian", logical "analysis". The comparison of the fragment *De la Sagesse* (1676) with this Cartesian concept is very interesting. Leibniz takes up the same model, but with the important distinction between "analysis of things" and "analysis of truth" mentioned in the previous section, i.e. he points to a possible gap between analysis of notions and truth based on simple notions.

and, more generally, as known since Euclid, no other (convex) regular polyhedra than the five constructed in the *Elements*[32].

Notwithstanding this latter formulation, another problem is that Descartes himself used notions like the maximum of all perfection in his demonstration of God's existence. Here we find again the problem mentioned in our introduction: if mathematics claims to give us the model of a general *ars inveniendi*, we should be aware that its use in metaphysics often implies infinitary arguments. So either we have to give up the dream of a *general* method, or we need to be able to build *from within mathematics* a model of analysis adapted to these kinds of situations. We are therefore in a position to better understand why Leibniz mentions this type of impossibility result to Elisabeth and why he does not take examples from his new Calculus. Not only are they directly connected to the default which Leibniz finds in Descartes demonstration of God's existence, but they also reveal the *intrinsic* difficulties of what Leibniz takes to be a Cartesian model of analysis[33].

2.2   The Quadrature of the Circle

The case of the quadrature of the circle is very similar to that of the "number of all numbers" in that Leibniz prepared, in 1675–1676, a treatise which was never sent for publication and which may likewise reveal the hidden motivation for many of his declarations. I refer to his *De quadratura arithmetica circuli ellipseos et hyperbolae*. Thanks to the work of Leibniz scholars, we now have at our disposal not only an edition of the treatise (E. Knobloch (1993)), but also the documents concerning the discovery of the arithmetical quadrature of the circle from the beginning of 1673 onwards[34].

My purpose is not to enter into the positive part of the construction, that is to say the discovery of the famous series for $\pi/4$ to which Leibniz eventually gave his name. I will only focus on its *pars destruens*, which is the result mentioned in the letter from 1678. It consists of a proof that no algebraic quadrature of the circle is possible. Once again, I will consider, for the sake of argument that Leibniz's proof is valid, even if we know that in order to be so it would require a conceptual apparatus which was not accessible before the nineteenth century.

---

[32] The last demonstration of Euclid's *Elements*, in Book XIII, establishes precisely the impossibility of constructing another one. Note that the passage devoted to the chiliogon in the *Meditationes* is in the same vein. Leibniz turns Descartes's example back on him, by noticing that the problem is not linked to the use of sensible imagination, but to the use of symbolic thinking in general (be it through words or through diagrams). Because we don't proceed to the analysis of the notions involve to its end, we therefore have no guarantee that this notion does not imply a contradiction. This becomes obvious if we replace polygons by regular polyhedra in the example.

[33] In contrast, the arguments taken from the new analysis simply indicate that Descartes's conception of "geometrical curves" was too narrow, without establishing clearly that his general "method" was responsible for this.

[34] They constitute a complete volume of the Academy Edition. See volume VII, 6: *Arithmetische Kreisquadratur* 1672–1676.

The proof is presented in the *Quadratura* as no less than the climax of the whole treatise (*velut coronis erit contemplationis hujus nostrae*)[35]. It consists in showing that there cannot be a "more geometrical" expression of the quadrature of the circle than the one given by an infinite series[36]. The argument is very simple: suppose, says Leibniz, that we possess an algebraic formula expressing the relationship of the tangent *t* to a given arc *a*, for example, a general equation of degree three in two variables in which some coefficients may be equal to zero (this would be the same as knowing how to "rectify" the circle, since the formula can be applied to an arc coinciding with the semi-circle). Since this equation holds for any arc, it would then imply that we also know the relationship between a tangent *t* and an arc smaller than the one which we chose, say an arc *b* of one eleventh of *a*. The equation in *b* is still the general equation (in our example a cubic). But this would mean that we can reduce the problem of the division of an angle into eleven parts to a third degree equation. This is absurd since Vieta's work *Sectiones angulatores* has demonstrated, according to Leibniz, that the division of an angle into a prime number of equal parts amounts to an equation which is not reducible to a lesser degree than the given prime number chosen[37]. Leibniz's argument, carried out for an equation of degree three, generalizes immediately: for any algebraic equation of degree *n*, we would just have to choose a prime number bigger than *n* to encounter the same contradiction.

This example is very important because it is situated, once again, on a Cartesian foundation. As Leibniz very nicely puts it in the letter to Elisabeth, even if some Cartesians might say that we cannot talk meaningfully of something which is impossible (because we can have no idea of what is impossible)[38], they would nonetheless need to talk meaningfully of impossible things in order to assess … their impossibility. The best example is given by the fact that Descartes had forcefully emphasized that the quadrature of the circle was impossible. Indeed such a statement was crucial in the delimitation of what was truly "geometrical" according to him. Leibniz is perfectly right to point out that this example shows the limitation of any Cartesian attempt to base knowledge on "clear and distinct" ideas, known in and of themselves. Indeed, either we do not have a "clear and distinct" idea of the quadrature of the circle, which means that we simply do not know anything about it, including whether or not it belongs to the realm of geometry. Or we do have a clear and distinct idea of it, but that would mean to have a "clear and distinct" idea

---

[35] It is the last of 51 proposition and its last words are: "*Impossibilis est ergo quadratura generalis sive constructio serviens pro data qualibet parte Hyperbolae aut Circuli adeoque et Ellipseos, quae magis geometrica sit, quam nostra est. Q.E.D.*" (Parmentier 2004, 354).

[36] "Geometrical" should be understood here in the Cartesian sense of "geometrical curve", i.e. one which can be expressed through a finite algebraic formula. Leibniz also has other arguments, which I shall not go into here, by means of which he is able to defend the fact that his particular series is better than others.

[37] This latter claim is not substantiated by Leibniz and one might doubt whether Vieta could be said to have produced a *demonstration* of this fact.

[38] Leibniz would agree with them on this point (which is why he carefully distinguishes between having a "notion" and having an "idea" of something).

of something which is ... impossible. This is the core of Leibniz's objection to Descartes's demonstration of God's existence: "clear and distinct" knowledge is not enough. When the analysis is not conducted to its end, it should be accompanied by a proof of the possibility of the notion involved in the reasoning.

Our two examples both proceed in the same direction: analysis of notions is not sufficient to ground perfect knowledge. We rarely reach truly simple notions and most of the times retain in our primitive terms hidden assumptions which reveal themselves only in the synthesis. This is why, as Leibniz sometimes puts it: "*il n'est pas en nostre pouvoir de faire des combinaisons à nostre fantaisie*"[39]. "The sum of all units" is a well-formulated mathematical sentence, just as the "algebraic quadrature of the circle" is, but the states of affairs to which these statements seem to refer do not exist. This is not just a matter of "evidence": in both cases, we have at our disposal *proofs* of this impossibility. But, as a consequence, we need to be able to reason clearly and distinctly about impossible objects in order to demonstrate their impossibility. This situation suffices to cast serious doubt on the initial program, or at least on the first and naive interpretation of it. It shows that the "combination" itself is not a transparent operation on concepts, if we did not reach perfectly simple elements. It cannot therefore be represented by a simple juxtaposition of letters. There is no purely logical incompatibility between the ingredients of a triangle, for example, the three sides which compose it, and the number ten. But when I "combine" ten triangles in order to construct a diamond, I stumble upon an impossibility which is caused by the special kind of synthesis involved in this "combination".

Moreover, this general situation shows that the reduction of synthesis to a mere computation (*Calculemus!*) would not be insured by the analysis of notions, as was at first expected. In this sense a *characteristica universalis* would be in exactly the same situation as classical Geometry (on which it is precisely modeled): the definitions, axioms and postulates of Euclid's *Elements* do not spare one the work of providing a demonstration that, for example, it is impossible for two circles to intersect at more than two points (prop. III, 10 of the *Elements*). When the proof is produced, one could then consider that this impossibility is a direct consequence of the primitive principles, but this is just retrospective reading of the kind of evidence that the proof, and only the proof, has provided. Nothing, in the definition of the circle and the principle of Euclidean geometry, prevents you from forming the "blind" combination: "the three points of intersection of two circles". This is a possibility that has to be excluded by the work of the proof itself and this proof is not reducible to a mere combination of notions (especially in the case mentioned, in which one has to rely on a *reductio ad absurdum*).

---

[39] GP V, 301; *Nouveaux Essais* III, 6, § 28. The continuation of the sentence being precisely: "*autrement on auroit droit de parler des Decaedres reguliers*" ("otherwise, one would have the right to speak about regular decahedra").

2.3   Prime Numbers

Now, it is also true that the "analysis of notions" contains some difficulties of its
own, i.e. which are not linked to hidden assumptions revealed by synthesis. The
most obvious is the fact that it seems particularly difficult to have a general method
for the detection of simple notions. More dramatically, and more interestingly, it is
not clear what a "simple notion" is. In this last section, I would like to show that
mathematical practice also played a role in emphasizing this fact .

As we have seen in the fragment entitled *Sur les premières propositions et les
premiers termes*, Leibniz compares the difficulty of finding basic notions with the
difficulty of recognizing prime numbers. This comparison comes as no surprise.
One of Leibniz' ideas in elaborating a first sketch of an algebra of thoughts, already
set out in the *De Arte Combinatoria*, was to represent basic notions by prime numbers and to represent the composition of notions through the way natural numbers
can be produced by the multiplication of primes. However, it is not always recognized that this had a ricochet effect on Leibniz's mathematical practice by orient-
ing his research towards a topic which was not considered as central at the time.
In contrast to Descartes in particular, Leibniz did not despise number theoretical
problems. He even devoted a lot of energy to their investigation during his stay in
Paris. Among these were the problems of ascertaining whether or not a number is
prime and where to find prime numbers in the scale of natural numbers[40]. These two
questions are now considered as basic problems in number theory, but in Leibniz's
day this was far from being the case, even amongst mathematicians interested in
problems of numbers[41].

Once again, my aim is not to enter into the technicalities of these questions, but
to emphasize their philosophical consequences. As Leibniz points out, we do not
possess a rule for describing exactly where to find the prime numbers on the scale
of natural numbers—something which is still true today. It is of interest to compare
this position with that of Pell who was one of the few mathematicians interested in
these questions at that time and whose work was well known to Leibniz. Pell devoted a lot of energy to the realization of huge tables of prime factors (up to 100,000).
This endeavor was incorporated into a philosophical program whose aim was to
find "prime truths" and to organize them into a systematic encyclopedia. This is
why Pell put a lot of emphasis on the fact that one should produce a "complete and
orderly enumeration" when producing a table of factors[42]. His program is very close
to Leibniz's and this is why it is particularly interesting to consider the differences
between them. Indeed, Leibniz did not consider that a "complete and orderly enumeration" of prime numbers was sufficient. In the passage from *Sur les premières*

---

[40] See *Ouverture nouvelle des nombres multiples, et des diviseurs des puissances*, January 1676
(A VII, 1, 576–578); *Figuram numerorum ordine dispositorum et punctatorum ut appareant qui
multipli qui primitivi* (A VII, 1, 579–581); *De numeris figuratis divisoribusque potestatum* (A VII,
1, 583–586); *De natura numerorum primorum et in genere multiplorum* (A VII, 1, 594–598).

[41] On the history of the study of prime numbers in the seventeenth century, see Bullynck (2010).

[42] Bullynck (2010) and Malcolm and Stedall (2005, 263–265).

propositions et les premiers termes, he even considers that mathematicians of his time were able to identify prime numbers only in an empirical way, "by trying to divide them". He then emphasizes the fact that on these matters what we need (and are still lacking) is a *universal rule*.

One could object that this is of no particular concern since we also have a simple additive model of natural numbers in which decomposition into units is always successful (as opposed to a multiplicative model in which decomposition involves prime numbers and the mysteries of their distribution). But that would be a very naive approach to the nature of mathematical knowledge. Indeed, the knowledge of a mathematical object, natural numbers in this case, is not exhausted by an adequate definition or our ability to derive well known facts such as "$2+2=4$". It also involves our ability to *find* new properties, that is to say, it involves the kind of questions which we are able to answer concerning it—a fact that we already encountered with the example of the intersection of two circles. Yet it so happens that most of the basic questions which one can address concerning natural numbers are at the same time very simple to formulate and very difficult to answer. Such is the case with some of the problems which interested Leibniz during his stay in Paris: to find a number which, after division by three given numbers, would give three given remainders; to find two squares equal to a given number; to find three numbers such that the sum of two of them is a square and the difference of two of them is also a square, etc[43]. It also happens that these kinds of problems involve more often than not questions of divisibility and of primality. In this sense, as is now well understood, questions concerning prime numbers and the way of recognizing them amongst natural numbers are not just a matter of describing natural numbers: they are also tools for use in the resolution of number-theoretic problems and therefore in gaining knowledge of their essential properties .

Nowadays, these remarks are trivial for mathematicians. But they were certainly not in Leibniz's time and he even may have been the first to realize their importance. They provide a general context to the parallel suggested between prime numbers and simple notions. The fact that we do not know how to find systematically prime numbers in the scale of natural numbers, except by tests and approximation, does not only concern the difficulty of finding simple elements—a fact which could always be interpreted as an indication of a temporary and purely contingent limitation of our knowledge. It also indicates that we do not master some *essential* mecanisms which enter into the solution of basic questions concerning numbers. This is a way of saying that we do not have "perfect knowledge" of some of the essential properties of the most simple, or at least apparently most simple, mathematical objects. At this point, we should keep in mind that natural numbers are the *only* example provided by the *Meditationes* of things which "come close" to notions capable of being fully analyzed.

This conclusion may possibly sound rather dramatic. What it means is that the true nature of arithmetic is to a large degree unknown to us. This is, in fact, explicitly

---

[43] This last problem occupied Leibniz so intensely that he made no less than thirty attempts at solving it—around four hundred pages in the Academy Edition. See Hofmann (1969).

stated by Leibniz, even in later texts: "*magna verae Arithmeticae pars hactenus sit ignorata*" (GM VII, 61)[44]. Hence the following dilemma: either we have to admit that our knowledge of numbers is not certain, because there is something in our understanding of their deep structure which lays hidden to us; or we must consider that our knowledge of numbers is certain, but partial. In this last conception, the problem is not that we lack a complete analysis—since we *do have* a complete analysis, or at least that which most resembles one—, but that a complete analysis *under one consideration* (additive model) can lead to a partial knowledge *under another one* (multiplicative model). This situation is still true: we possess very satisfactory axiomatic characterizations of natural numbers, but this does not mean that we really understand what a natural number is; this latter fact emerges clearly from the enormous difficulties lying in number-theoretic problems[45].

That this example belongs to the same line of argument against the Cartesian model of "method" as the one Leibniz presented in his letter of 1678 is made clear in a letter to Malebranche dating from the same period:

> I would like to know if your Mr. Prestet still works in analysis. I hope so, because he seems able in it. I recognize more and more the imperfection of the one we have at our disposal. For example, it provides no sure way to solve the problems in Diophant's Arithmetic [...]. Finally, I could write a book about the fields in which it does not succeed and where any Cartesian whosoever could not succeed without inventing some method beyond Descartes's method[46].

---

[44] See also this striking passage from the *Nouveaux Essais* about prime numbers: "C'est la multitude des considérations aussi qui fait que dans la science des nombres même il y a des difficultés très grandes, car on y cherche des abregés et on ne scait pas quelquesfois, si la nature en a dans ses replis pour le cas dont il s'agit. Par exemple, qu'y a t-il de plus simple en apparence que la notion du nombre primitif? c'est à dire du nombre entier indivisible par tout autre excepté par l'unité et par luy même. Cependant on cherche encor une marque positive et facile pour les reconnoistre certainement sans essayer tous les diviseurs primitifs, moindres que la racine quarrée du primitif donné. Il y a quantité de marques qui font connoistre sans beaucoup de calcul, que tel nombre n'est point primitif, mais on en demande une qui soit facile et qui fasse connoistre certainement qu'il est primitif quand il l'est" (*Nouveaux Essais* IV, 17, § 9, GP V, 470).

[45] The demonstration of Fermat's last theorem was certainly the most spectacular example of this fact in recent years, but there are many other examples in present day mathematics. One could mention that this is also something which was emphasized by the Bourbaki group in their structuralist manifesto, *L'architecture des mathématiques*: "[...] in certain theories (for example in the theory of Numbers), there exist many isolated results that up till now no one has been able to classify, nor connect in a satisfactory way with known structures" (Bourbaki 1950).

[46] "Je voudrois sçavoir si vostre M. Prestet continue à travailler dans l'analyse. Je le souhaite, parce qu'il y paroist propre. Je reconnois de plus en plus l'imperfection de celle que nous avons. Par exemple, elle ne donne pas un moyen seur pour resoudre les problemes de l'Arithmetique de Diophante [...] Enfin, je pourrois faire un livre des recherches où elle n'arrive point, et où quelque Cartesien que ce soit ne sçauroit arriver sans inventer quelque methode au delà de la methode de des Cartes. (Letter to Malebranche, January 1679; A II, 1, 677). See also A VI, 4, 2047 (1689) and *Nouveaux Essais* IV, 2, § 7: "On n'a pas encore trouvé l'analyse des nombres: Il arrive aussi que l'induction nous presente des verites dans les nombres et dans les figures dont on n'a pas encor decouvert la raison generale. Car il s'en faut beaucoup, qu'on soit parvenu à la perfection de l'Analyse en Geometrie et en nombres, comme plusieurs se sont imaginés sur les Gasconnades de quelques hommes excellens d'ailleurs, mais un peu trop prompts ou trop ambitieux" (GP V, 349).

Even if Leibniz did not mention Diophantine analysis to Elisabeth, it is clear from the preceding passage that it represented to his eyes another important example of the limitations of Descartes's "method".

Conclusion

The mathematical examples discussed in this paper contain in my opinion two major points of interest. First, they are *explicitly* mentioned by Leibniz when he comments on the relationship between mathematics and philosophy at the end of the 1670s and the beginning of the 1680s. As I have tried to show, they did indeed play a major role in his philosophical reflections during his stay in Paris and continued to remain important in his later thought[47]. Second, these mathematical examples provide a much more profound understanding of the doubts expressed in section II than those that first come to mind. Indeed it is very tempting to understand the difficulty related to the *analysis notionum* as being purely contingent and factual: we do not possess a complete analysis of human thoughts and if it were a feasible task it would certainly not be an easy one, especially for a single person. But as we saw through our examples, this is just one part of the story and perhaps the most superficial one. In fact, the real difficulty of "being simple" for notions is not only that it is difficult to achieve a complete analysis of notions, but that, *even if it were feasible*, it would leave untouched major difficulties which Leibniz encountered very early in his mathematical practice.

One of them is linked to the fact that a logical analysis offers no guarantee of not producing further contradictions if the analysis is not truly complete or, in other words, if the expected synthesis involves constraints which are not purely "logical". A regular decahedron is not in and of itself a contradictory notion: its inherent contradiction is attached to specific constraints characterizing the admissible constructions in Euclidean Geometry—in the same way that a triangle in which the sum of all angles is less than 180° is not in and of itself a contradictory notion, but may be incompatible with assumptions given by certain axioms. In other words, the problem is not only with the analysis of notions, but with the analysis of axioms (or, in Leibniz's parlance: "proof of axioms"). As we saw above, there is another type of difficulty attached to the same fact: as long as one begins with a *given* characterizations of domains of objects, even a complete analysis offers no guarantee of uncovering the "true nature" or "essence" of the objects under study. This is linked to the fact that an apparently complete analysis under one perspective could appear as incomplete under another one—this may be a reason why Leibniz says in the

---

[47] This situation has to be contrasted with the fact that Leibniz was also very explicit about some connections which he resisted making—although modern commentators tend to put a lot of emphasis on them. One famous example is given by the provocative declaration made to Masson in 1716: "The infinitesimal calculus is useful with respect to the application of mathematics to physics; *however, that is not how I claim to account for the nature of things*. For I consider infinitesimal quantities to be useful fictions" (GP VI, 629; transl. in Ariew & Garber, 230).

Meditationes that the analysis of numbers comes only close to a perfect example of complete analysis. Although one can easily recover the multiplicative structure of natural numbers from the additive structure, that does not give one a hint of the fact that the first perspective seems to be the accurate one to approach natural numbers. A testimony of this difficulty is the fact that problems easy to formulate in that domain still strongly resist to us.

In fact the difficulty is much deeper that might first be thought: one has simply *no way* to guarantee that the perspective adopted to engage the logical analysis is the accurate one. The only way to do so would be to already dispose of a complete analysis *under all possible (essential) perspectives* so as to be sure that the characterization under study is not only attached to essential properties, but also truly complete (i.e. entailing *all* the essential properties characterizing the objects under study). This latter point touches a very profound difficulty in any kind of "analytical" program. Indeed, there is no notion of "complete" analysis if we do not already have *at hand* an accurate characterization of the notions involved—at least a *complete* list of the prerequisites constituting the "essential" properties of a given concept. But how can we have at hand an accurate delimitation of the notions if we do not *already* have accurate knowledge of them, that is to say ... a complete analysis of them? The example of natural numbers is very interesting in this regard. We certainly do have, according to Leibniz, a "clear and distinct" notion of them, but this does not mean that our knowledge of them is *adequate*. All we can say, and all that Leibniz says, is that it is what "comes close" to it. But we also have to admit, and Leibniz had no difficulty in admitting this fact, that a large part of the science of numbers is still *hidden to us*.

By assembling the different pieces which we have encountered so far, one might have the impression that Leibniz's "grand program" was just a youthful dream and that it could not survive the kind of difficulties which he began to face very early in his mathematical practice. But this would be to go too far. What Leibniz actually concluded was more likely that it was possible to reach *perfect* knowledge without having completed the analysis of notions. How is this possible? How can we be sure, for example, that our knowledge does not imply a hidden contradiction if we are not able to pursue the analysis of notions to its end? Leibniz is very clear about the different solutions which he proposed to these difficulties. In the *Meditationes* he sketches two other strategies alongside complete analysis to insure the possibility of the notions involved: experience and causal definitions[48]. Here is not the place to describe these in detail, but we should at least take note that they have not received much attention amongst scholars studying Leibniz's philosophy of mathematics. This is particularly true of the role of experience[49]. Another strategy presented in the

---

[48] A VI, 4, 489.

[49] One can remember here the striking formulation of "La vraie méthode" (1677): "certain experiments are *always* necessary to serve as a basis for reasoning" (A VI, 4, 3. My emphasis). One could object that, according to the *Meditationes*, experience can only serve as basis for an *a posteriori* proof. But the fact that mathematical truths are *a priori* does not exclude in and of itself the validity of *a posteriori* proof of their possibility (especially if we don't have access to complete analysis of these notions).

De Elementis cogitandi, is to reduce the demonstration to "identicals". According to this, "a demonstration is perfect as soon as one can reach identicals, which can happen even though everything is not analysed" (A VI, 3, 504).

Although it is not clear how the reduction to "identicals" could keep us from the danger of hidden contradictions, this strategy was certainly *practiced* by Leibniz and it is in fact a very good example of how reflection on general method led him to new considerations *in mathematics*. The catalog of basic relations which Leibniz produced for geometry in his numerous fragments devoted to the realization of a "characteristica geometrica" are what we now call "equivalence relations" and often mentioned by him amongst "identicals"[50]. We have here a beautiful example of a mathematical practice which seems to be derived from the reevaluation of the *analysis notionum* and it is important to note that the relationship between mathematics and philosophy was certainly not one-way in Leibniz's thought[51]. But there is more to be said about this last strategy. The fact that, in mathematics, one can stop the logical analysis with equivalence relations indicates a specificity of this form of knowledge. Mathematical objects, in contrast to "real" objects, are indeed characterized by a kind of indiscernibility. But as soon as one realizes the difference between two types of notions ("complete" and "incomplete"), one has also to admit that the logical analysis can no longer work according to the same model in metaphysics and in mathematics. This is a good point upon which to close our investigation: the *analysis notionum* splits here into two types of tasks, depending on the kind of notions involved and the kind of simplicity which is reachable in it. This case indicates once again the way in which mathematics and philosophy really did interact in Leibniz's thought in a constant and complex dialog.

Acknowledgements I would like to thank Philip Beeley, Norma Goethe, and an anonymous referee for their helpful comments and corrections on this paper.

References

Bourbaki, N. 1950. The architecture of mathematics. *The American Mathematical Monthly* 57:221–232.

Bullynck, M. 2010. A history of factor tables with notes on the birth of number theory 1657–1817. *Revue d'histoire des mathématiques* 16:133–216.

Couturat, L. 1901. *La Logique de Leibniz d'après des documents inédits*. Paris: Alcan.

Hofmann, J. E. 1969. Leibniz und Ozanams Problem. *Studia Leibnitiana* 1:103–126.

Hofmann, J. E. 1974. *Leibniz in Paris (1672–1676). His growth to mathematical maturity*. Cambridge: Cambridge University Press.

Leibniz, G. W. 1951. *Leibniz selections* (ed. and transl. by Philip P. Wiener). New York: Scribner's.

---

[50] See the list of "identicals" given to Conring in the letter from March 1678 (A II, 1, 599) or the one given ten years after at the beginning of the *Principa logico-metaphysica* (A VI, 4, 1645; 1689).

[51] On this issue, see Rabouin (2013), "*Analytica Generalissima Humanorum Cognitionum*. Some reflections on the relationship between logical and mathematical analysis in Leibniz".

Leibniz, G. W. 1981. *New essays on human understanding*. Trans: Peter Remnant and Jonathan Bennett. Cambridge: Cambridge University Press.

Leibniz, G. W. 1989a. *Philosophical essays*. Trans: D. Garber and R. Ariew. Indianapolis: Hackett.

Leibniz, G. W. 1989b. *Philosophical papers and letters*. 2nd ed. Trans: Leroy E. Loemker. Dordrecht: D. Reidel.

Leibniz, G. W. 1993. *De quadratura arithmetica circuli ellipseos et hyperbolae cujus corollarium est trigonometria sine tabulis*, critical edition and commentary by Eberhard Knobloch. Abhandlungen der Akademie der Wissenschaften in Göttingen, Mathematisch-physikalische Klasse 3. Folge, Nr. 43, Göttingen: Vandenhoeck & Ruprecht, 1993; French translation: *Quadrature arithmétique du cercle, de l'ellipse et de l'hyperbole et la trigonométrie sans tables trigonométriques qui en est le corollaire*; introd., trad. et notes de Marc Parmentier, Paris: Vrin, 2004 (coll. "Mathesis").

MacDonald Ross, G. 1990. Are there real infinitesimals in Leibniz's metaphysics? In *L'infinito in Leibniz: Problemi e terminologia,* ed. A. Lamarra, 125–141. Rome: Edizioni dell'Ateneo.

Malcolm, N., and J. Stedall. 2005. *John Pell (1611–1685) and his correspondence with Sir Charles Cavendish: The mental world of an early modern mathematician*. Oxford: Oxford University Press.

Picon, M. 2003. Vers la doctrine de l'entendement en abrégé: éléments pour une généalogie des *Meditationes de cognitione, veritate, et ideis. Studia Leibnitiana* 35:102–132.

Pombo, O. 1987. *Leibniz and the problem of a universal language*. Münster: Nodus Publikationen.

Rabouin, D. 2013. *Analytica Generalissima Humanorum Cognitionum*. Some reflections on the relationship between logical and mathematical analysis in Leibniz. *Studia Leibnitiana* 45: 109–130.

Rossi, P. 2006. *Clavis universalis. Arti della memoria e logica combinatoria da Lullo a Leibniz*, Milano-Napoli: Ricciardi, 1960; english transl. and intr. by Stephen Clucas, *Logic and the Art of Memory. The Quest for a Universal Language*. London: Continuum Press.

Sasaki, C. 2003. Descartes's *mathematical thought*. Dordrecht: Klüwer.

Part II
Mathematical Reflections

Leibniz's Mathematical and Philosophical Analysis of Time

Emily R. Grosholz

Leibniz believed that mathematics has a special place in the human search for wisdom, knowledge of the "most sublime principles of order and perfection," because the things of mathematics are so determinate, and exhibit their determinate inter- relations so clearly. However, the proper use of mathematics requires careful philosophical reflection. The reason why materialism has seemed attractive to serious thinkers, he argues in the *Tentamen Anagogicum* (1696), is because it lends itself well to mathematical representation, and thus to calculation and rigorous inference.[1] However, we should not over-estimate the extent to which the material world lends itself to mathematics, for all mathematical 'models' are a finitary representation of an infinitary reality; and we should not forget that other aspects of reality also lend themselves similarly to mathematization. The materialist illusion is not only a mathematical mistake (which should be addressed by yet more mathematics) but also a metaphysical mistake. The alleged materialist universe is a mirage, for it violates the principle of sufficient reason, which along with the principle of contradiction governs the created world; it is thus after all not thinkable, like the mirage of the 'greatest speed.' The world's beings are not only material, but thoroughly sentient and endowed with force or conatus, a striving for perfection; and in that striving they express their Maker, as well as the intelligibility for which mathematics is apt.

---

[1] Leibniz, G. W. *Philosophische Schriften*, ed. C. I. Gerhardt, Vol. VII, pp. 270–279. Abbreviated hereafter as 'GP' with reference to volume and page number.

E. R. Grosholz (✉)
Department of Philosophy, Pennsylvania
State University, University Park, PA, USA
e-mail: erg2@psu.edu

Laboratoire SPHERE, UMR 7219,
CNRS—Université Paris Diderot, Paris, France

© Springer Netherlands 2015
N. B. Goethe et al. (eds.), *G.W. Leibniz, Interrelations between Mathematics and Philosophy,* Archimedes 41, DOI 10.1007/978-94-017-9664-4_4

1 Leibniz on Method

Leibniz writes that "the ancients who recognized nothing in the universe but a concourse of corpuscles", as well as the modern philosophers who are inspired by them, find materialism plausible,

> because they believe that they need to use only mathematical principles, without having any need either for metaphysical principles, which they treat as illusory, or for principles of the good, which they reduce to human morals; as if perfection and the good were only a particular result of our thinking and not to be found in universal nature... It is rather easy to fall into this error, especially when one's thinking stops at what imagination alone can supply, namely, at magnitudes and figures and their modifications. But when one pushes forward his inquiry after reasons, it is found that the laws of motion cannot be explained through purely geometric principles or by imagination alone. (GP VII, 271)[2]

Moreover, he adds, there is no reason to suppose that other phenomena which in that era had eluded mathematical formulation (he mentions light, weight, and elastic force) will not sooner or later prove to lie within the expressive powers of mathematics. But all such representation will be provisional, because while finitary models can express the infinitary things of nature well, they can never express them completely; and the formulation of increasingly accurate stages of representation must be governed, like nature itself, by the two great principles of contradiction and sufficient reason.

Leibniz recognizes that different sciences require different methodologies, but no matter what special features different domains exhibit, he believes that all scientific investigation must move between mathematics and metaphysics. Mechanics, in particular, is best viewed as a middle term between mathematics and metaphysics, and so too Leibniz's account of time. Of all the parameters involved in mechanics, time is the least tied to any specific content, even though it presents a determinate topic for scientific investigation. Thus a closer look at Leibniz's account of time presents an especially 'pure' version of the interaction of mathematics and philosophy in the service of progressive knowledge.

As Yvon Belaval, Gilles-Gaston Granger, François Duchesneau, and Daniel Garber have variously argued on the basis of a wide range of texts, Leibniz's novel conception of scientific method has two dimensions (Belaval 1960; Granger 1981; Duchesneau 1993; Garber 2009). His account of method is informed by that of Bacon and Descartes, but diverges from both in significant ways and combines aspects of each. He borrows from Bacon the project of collecting empirical samples from the laboratory and field, inductively, and compiling tables, taxonomies and encyclopediae, always with the expectation of discovering harmonies and analogies, deeper systematic organization in the things of nature. He borrows from

---

[2] "Parce qu'ils croyent de n'avoir à employer que des principes de mathematique, sans avoir besoin ny de ceux de metaphysique qu'ils traitent de chimeres, ny de ceux du bien qu'ils renvoyent à la morale des hommes, comme si la perfection et le bien n'estoient qu'un effect particulier de nos pensées, sans se trouver dans la nature universelle... il est assez aisé de tomber dans cette erreur, et par tout quand on s'arreste en meditant à ce que l'imagination seule peut fournir, c'est à dire aux grandeurs et figures, et à leurs modifications. Mais quand on pousse la recherche des raisons, il se trouve que les loix du mouvement ne scauroient estre expliquées par des principes purement geometriques, ou de la seule imagination." (GP VII, 271).

Descartes the assurance that the indefinite presentations of sense can be associated with precise mathematical concepts, and thus by analogy be re-organized as ordered series, which can then be subject to deductive inference.

In the *Tentamen Anagogicum*, Leibniz mentions the use of geometry in the "analysis of the laws of nature", and goes on in that essay to develop the ideas of Fermat, Descartes, and Snell in optics using a series of geometrical diagrams, as well as the ideas of maximal and minimal quantities developed in his infinitesimal calculus. In an earlier, more general essay, "Projet d'un art d'inventer" (1686), he invokes arithmetic as a source of formulations apt for analysis considered as the art of invention, "which would have the same effect in other subject matters, like that which algebra has on arithmetic. I have even found an astonishing thing, which is that one can represent all kinds of truths and inferences by means of numbers." (C 175) [3] The idea is to locate nominal definitions, involving a finite number of requisites, and then reason on the basis of them:

> I found that there are certain primitive terms —if not absolutely primitive then at least primitive for us—which once having been consituted, all our reasonings could be made determinate in the same way as arithmetical calculations; and even in the case of those reasonings where the data, or given conditions, don't suffice to determine the question completely, one could nevertheless determine [metaphysically] mathematically the degree of probability. (C 176)[4]

The clarity and determinacy of mathematical things is crucial to this method of analysis. "The only way to improve our reasonings is to make them as salient as those of mathematicians, so that one can spot an error clearly and quickly, and when there is a dispute, one need only say: let us compute, without further ado, to see who is right." (C 176)[5]

Early modern mechanics begins by exploiting an already existing trove of empirical records, the precise tables left by centuries of astronomers tracking the movements of the moon, the planets, certain stars and the named constellations which culminate in the careful data of Tycho Brahe, so important to Kepler, and which are soon thereafter improved by the measurements of astronomers equipped with telescopes. Happily for human science, the solar system is both an exemplary mechanical system (just a few moving parts, isolated, and so almost closed despite

---

[3] "qui feroit quelque chose de semblable en d'autres matieres, à ce que l'Algebre fait dans les Nombres. J'ay même trouvé une chose estonnante, c'est qu'on peut representer par les Nombres, toutes sortes de verités et consequences." (Leibniz, G. W. *Opuscules et fragments inédits*. Ed. L. Couturat. Hildesheim: Georg Olms, p. 175. Abbreviated hereafter as 'C' with reference to page number).

[4] "Je trouva donc qu'il y a des certains Termes primitifs si non absolument, au moins à nostre egard, les quels estant constitués, tous les raisonnements se pourroient determiner à la façon des nombres et meme à l'egard de ceux ou les circonstances données, ou data, ne suffisent pas à la determination de la question, on pourroit neantmoins determiner [Metaphysiquement] mathematiquement le degré de la probabilité." (C 176) (Couturat indicates by brackets a word or phrase that Leibniz has crossed out.).

[5] "L'unique moyen de redresser nos raisonnemens est de les rendre aussi sensibles que le sont ceux des Mathematiciens, en sorte qu'on puisse trouver son erreur à veue d'oeil, et quand il y a des disputes entre les gens, on puisse dire seulement: contons, sans autre ceremonie, pour voir lequel a raison." (C 176).

the occasional comet) and a very precise clock; so its study richly repaid the efforts of early modern physicists.

How shall these two occupations, empirical compilation and theoretical analysis, be combined? Leibniz calls on metaphysics, in particular the principle of sufficient reason in the guise of the principle of continuity, to regulate a science that must be (due to the infinite complexity of individual substances) both empirical and rationalist. The correlation of precise empirical description with the abstract conception of science *more geometrico* is guaranteed by the thoroughgoing intelligibility and perfection of the created world, and encourages us to work out our sciences through successive stages, moving back and forth between a concrete taxonomy and abstract systematization. Empirical research furnishes nominalist definitions—finite lists of requisites for the thing defined—which can set up the possibility of provisionally correct deductions, though every such definition due to its finitude can be corrected and amplified; mathematics provides the rule of the series.

At the beginning of Chap. 6, "La philosophie de l'histoire" of his book *Leibniz historien*, Louis Davillé writes:

> From the metaphysical point of view, Leibniz, contemplating together the diversity and uniformity of things and beings, also follows two opposed principles, recognized earlier by scholastic philosophers, the principle of individuation and the principle of analogy, which he expresses by two phrases, in French: "l'individualité enveloppe l'infini" and "c'est tout comme ici." But this is only an appearance. Always seeking to reconcile opposites, he unites these two points of view in "la conception d'un développement à la fois spontané et régulier des êtres,"[6] through the contemplation of the universal harmony, principle of things persisting in diversity balanced by identity. This powerful and original synthesis he calls the law of continuity … The notion of continuity plays a leading role in Leibniz's philosophy, differentiating it sharply from that of Descartes. One might call the law of continuity the 'general method' of Leibniz, and this expression doesn't seem to be an exaggeration. (Davillé 1909, pp. 667–68)

Davillé notes three formulations of the principle of continuity: (1) Time and space are divisible to infinity. (2) The order of the input terms ('principes') is expressed in the order of the output values ('consequences') and vice versa. (I use the anachronistic vocabulary of functions here, to capture the generality of Leibniz's words.) This principle, 'of harmony', is a corollary of the principle of reason. It can also be understood as the principle of induction, that the cause can always be retrieved from the effect; the principle of differentials (ratios between finite magnitudes persist even when the magnitudes are reduced to infinitesimals, as in the 'characteristic triangle') ; and the principle of analogy. (3) Change never occurs in jumps, but always by degrees. Leibniz also calls this the principle of transition; like the principle of the identity of indiscernibles, Leibniz deduces it from the principle of sufficient reason. The principle of continuity, taken as a principle governing *history*, corresponds to a conception of historical evolution, slow and successive change due to natural and immanent causes. (Davillé 1909, pp. 668–670)

This model of scientific inquiry accords very well with Leibniz's own investigations into mechanics and planetary motion, and so too his mathematical-metaphysical

---

[6] Davillé quotes Delbos in this context. See V. Delbos. *La philosophie pratique de Kant*. Paris: Alcan, 1905, p. 264.

account of time. Given the subtlety of his conception of method, I will argue that his account of time is deeper and more multivalent than that of Newton, which explains why it has proved to be more suggestive for physicists in succeeding eras and especially during the last century.

2   Descartes and Newton

Descartes' definition of motion in the *Principles* is "the transfer of one piece of matter, or one body, from the vicinity of those bodies which are in immediate contact with it, and which are regarded as being at rest, to the vicinity of other bodies." (AT VIII, 53).[7] Thus motion and rest can be interpreted only as a difference in velocity or acceleration established with respect to a reference frame of other bodies; no absolute determination of motion or rest is possible. This definition of motion and rest is so radically relativistic that, strictly speaking, the Cartesian observer, by choosing different reference frames, may not only shift from judging that a given particle is at rest to judging that it is in inertial motion (rectilinear motion at a constant speed), but also to judging that its trajectory should be considered accelerated (and perhaps curvilinear). Descartes himself never seems to have considered this consequence of his relativism, nor its inconsistency with his invocation of inertial motion in the first two rules of motion given at the beginning of the *Principles*. Perhaps the inconsistency escaped his notice because in his mechanics there is no accelerated motion: the inherent motion of corpuscles is rectilinear and constant in speed (that is, inertial) and the transfer of momenta (defined for each contributing corpuscle as bulk times constant speed) in a collision is instantaneous. His mechanics is thus undynamical and atemporal; its laws are not only time-reversal invariant, they do not involve time as an independent variable: nothing in Descartes' mechanics varies continuously with respect to time.

Newton, however, saw and criticized this outcome, precisely because it entails that Descartes is not entitled to his own definition of inertial motion. In *De Gravitatione* (unpublished in his lifetime) he argues that since in Cartesian vortex mechanics all bodies are constantly shifting their relative positions with time, "Cartesian motion is not motion, for it has not velocity, nor definition, and there is no space or distance traversed by it. So it is necessary that the definition of places, and hence of local motion, be referred to some motionless thing such as extension alone or space in so far as it is seen to be truly distinct from bodies" (Newton 1962, p. 131). That is, Descartes cannot give empirical procedures in his mechanics that allow him to distinguish inertial motion from accelerated motion.

Newton responds with his well known thought experiment about the revolving bucket, arguing that the presence of forces is the sign of true (accelerated) motion; forces are real and measurable. But he goes beyond that claim: in Book III of the *Principia*, he writes,

---

[7] Descartes (1964–1974).

Hypothesis I: The center of the system of the world is at rest.
Proposition 11, Theorem 11: The common center of gravity of the earth, the sun, and all the
planets is at rest. (Newton 1999, p. 816)

Taken together, these claims offer an absolutist conception of space that makes
not only accelerated motion, but even uniform motion, definable with respect to a
Euclidean space that has been provided with a centre and axes. By countering so
strongly Descartes' relativism and subsequent loss of the distinction between inertial motion and accelerated (straight or curvilinear) motion, Newton has sacrificed
the equivalence of inertial reference frames and thus his own first law. He has also
postulated a spatio-temporal structure that cannot be empirically verified, a set of
Cartesian coordinates for the Euclidean space of his planetary mechanics, which
violates his methodological principle of not invoking merely metaphysical hypotheses. Newton is not entitled to the equivalence of rest and inertial motion, which
is just as essential to his system as Descartes' concept of inertial motion is to his
system. (Grosholz 2011)

3 Leibnizian Time

Leibniz acknowledged but was not troubled by the consequences of Descartes' relativism, and extended it to time. Thus in a commentary on the *Principles*, "Critical
Thoughts on the General Part of the Principles of Descartes" (unpublished in his
lifetime), Leibniz writes about *Principles* II, Articles 25 and 26:

If motion is nothing but the change of contact or of immediate vicinity, it follows that we
can never define which thing is moved. For just as the same phenomena may be interpreted
by different hypotheses in astronomy, so it will always be possible to attribute the real
motion to either one or the other of the two bodies which change their mutual vicinity or
position. Hence, since one of them is arbitrarily chosen to be at rest or moving at a given
rate in a given line, we may define geometrically what motion or rest is to be ascribed to
the other, so as to produce the given phenomena. Hence if there is nothing more in motion
that this reciprocal change, it follows that there is no reason in nature to ascribe motion to
one thing rather than to others. The consequence of this will be that there is no real motion.
(GP IV, 369)[8]

This is just what Newton says! But for Leibniz, it is not a problem, certainly not a
problem to be banished by postulating absolute space and time as the arena for motion. Rather, he makes the following claim: "Thus, in order to say that something
is moving, we will require not only that it change its position with respect to other

---

[8] "Si motus nihil aliud est quam mutatio contactus seu viciniae immediatae, sequitur nunquam
posse definiri, quaenam res moveatur. Ut enim in Astronomicis eadem phaenomena diversis hy-
pothesibus praestantur, ita semper licebit, motum realem vel uni vel alteri eorum tribuere quae
viciniam aut situm inter se mutant; adeo ut uno ex ipsis pro arbitrio electo, tanquam quiescente, aut
data ratione in data linea moto geometrice definiri queat, quid motus quietisve reliquis tribuendum
sit, ut data phaenomena prodeant. Unde si nihil aliud inest in motu, quam haec respectiva mutatio,
sequitur nullam in natura rationem dari cur uni rei potius quam aliis ascribi motum oporteat. Cujus
consequens erit, motum realem esse nullum." (GP IV, 369).

things but also that there be within itself a cause of change, a force, an action."[9] Newton proposes that whenever acceleration occurs, it is due to the action of forces; Leibniz proposes that whenever any motion occurs, it is due to the action of forces. This doesn't mean that he has reverted to Aristotelianism, but is instead an expression of his pan-animism. What Leibniz means by force is not Newtonian force, but something more like energy, internal to the body. Leibniz believes that no body is ever truly at rest, for all bodies are ensouled: motion thus becomes an expression of *conatus*, as individual substances jostle each other for a place within the Cartesian plenum at all times. (GP IV, 354–392)

In this picture of the universe, we see the principle of sufficient reason at work, fashioning Lebniz's mechanics along with mathematics. The universe must be a plenum, and the individual substances in that plenum are jostling each other in an effort to attain perfection: everything strives. Indeed for Leibniz even unactualized possibles strive: essences strive for existence. In the realm of ideas, this striving sorts ideas out into an infinity of possible worlds, and (with the beneficent cooperation of God) precipitates one world into creation; in the created world, it induces vortical motion in the plenum as well as temporality. Time is the expression of the incompatibility of things; because creation involves plurality, mentality, and mutual limitation, all things are active, passive and intentional. This is the best of all possible worlds because it is continually becoming more perfect, on into the infinite open future: creation is a continuous temporal process. In the law of the series, the independent variable is always time. Thus matter is not merely extended, but involves resistance and action; and it develops: Leibniz's science will also be a natural history.

Having invented a supple and powerful notation for his version of the infinitesimal calculus during his sojourn in Paris (1672–1676), Leibniz proceeded to work out a theory and practice of differential equations, in which the dependence of different forms of accelerated motion on time could be clearly expressed by the term 'dt'. One application of this method was to planetary motion. While in Vienna on his way to Rome in 1688, Leibniz read Newton's *Principia*, took extensive notes and then wrote a series of papers that culminated in the *Tentamen de Motuum Coelestium Causis* (*Acta Eruditorum*, Feb. 1689), where he proposed differential equations that would characterize planetary motion. Leibniz combined Cartesian vortex theory with Newton's reformulation of Kepler's laws, locating the planets in 'fluid orbs' rather than empty space, in order to derive the laws governing central forces while avoiding the problem of action at a distance. Whereas Newton calculates the deviation from the tangent to the curve, Leibniz expresses the situation with a single differential equation, by calculating the variation of the distance from the center, comparing the distances at different times by a rotation of the radius. The upshot of his calculation is that the effect of gravity is $[(2h^2)/(ar^2)]\,dt^2$, so that the 'solicitation of gravity' (conceptualized in Cartesian terms as the action of a vortex) is inversely proportional to the square of the distance, which was of course the result Leibniz was trying to reproduce. (Aiton 1985, Chap. 6; Bertoloni Meli 1993, Chap. 4)

---

[9] "Itaque ad hoc, ut moveri aliquid dicatur, requiremus non tantum ut mutet situm respectu aliorum, sed etiam ut causa mutationis, vis, actio, sit in ipso." (GP IV, 369).

4   Leibnizian Relationalism

For Leibniz, space is the expression in the created world of the logical order of compossibility among individual substances, and time is the logical order of incompatability among individual substances.[10] Thus, space and time only come into being with the creation of this material universe, the best of all possible worlds, and have only a secondary ontological status, because they are constituted as relational structures of the things with primary ontological status, individual substances. This is the basis of Leibniz's relationalism; but we must recall that his relationalism is deployed on the basis of a method which is two-tiered, both mathematical (seeking a precise mathematical correlate for the law of the series) and metaphysical while at the same time empirical (examining and tabulating evidence in an ongoing search for the systematic organization of things). The true scientist will find ways to put the mutual adjustment of nominalistic form with the investigation of the infinitely complex, infinitely ordered world of individual substances, in the service of the progress of knowledge; this process requires both mathematics and metaphysics.

To correlate time with precise mathematical concepts, Leibniz chooses as the correct representation the straight Euclidean line, endowed with directionality by Descartes' analytic geometry, which assigns positive and negative numbers—real numbers we would say—to the line. In some texts, it appears that Leibniz holds time to be a half-line, given what he writes to Clarke in the fifth letter of the Leibniz-Clarke correspondence (GP VII, 389–420). Since this is the best of all possible worlds, created by God, the universe must constantly increase in perfection, and so has a temporal beginning point but no end. Thus it is metaphysically important that the number-line is both geometrical and arithmetical. As arithmetical, it expresses the fact that time is asymmetric; time may be counted out in units, like seconds or years, and the numbers increase in a unidirectional order without bound to infinity. The asymmetry of time follows from the metaphysical ground that everything strives. As geometrical, the number-line expresses the fact that time is a continuum; units of time like seconds are not atoms, but conventionally established, constant measures of time, as the inch is a measure of continuous length. An instant is only the marker of a boundary of a stretch of time, not what time is composed of; we misunderstand what an instant is, Leibniz observes, if we conceive of it as an atom of time. Time must be both measured and counted.

This duality of time is not however without conundrums. Analysis in arithmetic leads us to the unit; but in geometry it leads us to the point. Whole numbers are composed of units, but lines are bounded by points, not composed of them; Cartesian reductionism is useful as an approach to arithmetic, but not to geometry. In a letter to Louis Bourguet, written just before the correspondence with Clarke, in August 1715, Leibniz writes,

> As for the nature of succession, where you seem to hold that we must think of a first, fundamental instant, just as unity is the foundation of numbers and the point is the foundation of

---

[10] See, for example, GP II, 248–53.

extension, I could reply to this that the instant is indeed the foundation of time but that since there is no one point whatsoever in nature which is fundamental with respect to all other points and which is therefore the seat of God, so to speak, I likewise see no necessity whatever of conceiving a primary instant. I admit, however, that there is this difference between instants and points—one point of the universe has no advantage of priority over another, while a preceding instant always has the advantage of priority, not merely in time but in nature, over following instants. But this does not make it necessary for there to be a first instant. There is involved here the difference between the analysis of necessities and the analysis of contingents. The analysis of necessities, which is that of essences, proceeds *from the posterior by nature to the prior by nature*, and it is in this sense that numbers are analyzed into unities. But in contingents or existents, this analysis *from the posterior by nature to the prior by nature* proceeds to infinity without ever being reduced to primitive elements. Thus the analogy of numbers to instants does not at all apply here. It is true that the concept of number is finally resolvable into the concept of unity, which is not further analyzable and can be considered the primitive number. But it does not follow that the concepts of different instants can be resolved finally into a primitive instant. (GP III, 581–582)[11]

The analysis of time requires the scientist to proceed both by the analysis of contingents, using the line whose continuity is the best expression mathematics provides for infinite complexity; and by the analysis of necessities, using the natural numbers whose linear ordering and asymmetry is the best mathematical expression of irrevocability. Leibniz goes on to observe that the use of mathematics does not solve the metaphysical question whether time has a beginning, which leads one to suppose that more metaphysics and more empirical research are required. He writes:

Yet I do not venture to deny that there may be a first instant. Two hypotheses can be formed—one that nature is always equally perfect, the other that it always increases in perfection. If it is always equally perfect, though in variable ways, it is more probable that it had no beginning. But if it always increases in perfection (assuming that it is impossible to give its whole perfection at once), there would still be two ways of explaining the matter, namely, by the ordinates of the hyperbola B or by that of the triangle C.[12]

---

[11] "Pour ce qui est de la succession, où vous semblés juger, Monsieur, qu'il faut concevoir un premier instant fondamental, comme l'unité est le fondement des nombres, et comme le point est aussi le fondement de l'etendue: à cela je pourrois reponre, que l'instant est aussi le fondement du temps, mais comme il n'y a point de point dans la nature, qui soit fondamental à l'egard de tous les autres points, et pour ainsi dire le siege de Dieu, de meme je ne vois point qu'il soit necessaire de concevoir un instant principal. J'avoue cependant qu'il y a cette difference entre les instans et les points, qu'un point de l'Univers n'a point l'avantage de priorité de nature sur l'autre, au lieu que l'instant precedent a tousjours l'avantage de priorité non seulement de temps, mais encor de nature sur l'instant suivant. Mais il n'est point necessaire pour cela qu'il y ait un premier instant. Il y a de la difference en cela entre l'analyse des necessaires, et l'analyse des contingens: l'analyse des necessaires, qui est celle des essences, allant *a natura posterioribus ad natura priora*, se termine dans les notions primitives, et c'est ainsi que les nombres se resolvent en unités. Mais dans les contingens ou existences cette analyse *a natura posterioribus ad natura priora* va à l'infini, sans qu'on puisse jamais la reduire à des elemens primitifs. Ainsi l'analogie des nombres aux instans ne procede point icy. Il est vray que la notion des nombres est resoluble enfin dans la notion de l'unité qui n'est plus resoluble, et qu'on peut considerer comme le nombre primitif. Mais il ne s'ensuit point que les notions des differens instans se resolvent enfin dans un instant primitif. " (GP III, 581–582).

[12] "Cependant je n'ose point nier qu'il y ait eu un instant premier. On peut former deux hypoteses, l'une que la nature est tousjours egalement parfaite, l'autre qu'elle croit tousjours en perfection. Si elle est tousjours egalement parfait, mais variablement, il est plus vraisemblable qu'il n'y ait

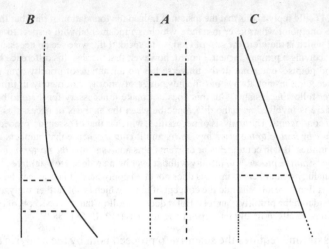

Fig. 1 Letter from Leibniz to Bourguet, 5 August 1715, GP III, p. 582

Here Leibniz gives the diagrams reproduced in Fig. 1.

His explanation of these diagrams shows that, despite what he would shortly write to Clarke, he was perhaps not convinced that time has a beginning:

> According to the hypothesis of the hyperbola, there would be no beginning, and the instants or states of the world would have been increasing in perfection from all eternity. But, according to the hypothesis of the triangle, there would have been a beginning. The hypothesis of equal perfection would be that of rectangle A. I do not yet see any way of demonstrating by pure reason which of these we should choose. But though the state of the world could never be absolutely perfect at any particular instant whatever according to the hypothesis of increase, nevertheless the whole actual sequence would always be the most perfect of all possible sequences, because God always chooses the best possible. (GP III, 582−83)[13]

In any case, Leibniz's conception of method requires that time be investigated not solely by pure reason or pure mathematics, which he admits here to being inconclusive; time must also be investigated empirically. It must be considered as the relational structure of the individual substances that exist, insofar as they are not logically compatible with each other. This means that we may have to revisit the formal structures we have just been discussing, in light of what we discover about

point de commencement. Mais si elle croissoit tousjours en perfection (supposé qu'il ne soit point possible de luy donner toute la perfection tout à la fois) la chose se pourroit encor expliquer de deux façons, savoir par les ordonnées de l'Hyperbole B ou par celle du triangle C." (GP III, 582).

[13] "Suivant l'hypothese de l'Hyperbole, il n'y auroit point de commencement, et les instans ou etats du Monde seroient crûs en perfection depuis toute l'eternité; mais suivant l'hypothese du Triangle, il y auroit eu un commencement. L'hypothese de la perfection egale seroit celle d'un Rectangle A. Je ne vois pas encor le moyen de faire voir demonstrativement ce qu'on doit choisir par la pure raison. Cependant quoyque suivant l'hypothese de l'accroissement, l'etat du Monde ne pourroit jamais etre parfait absolument, etant pris dans quelque instant que ce soit; neanmoins toute la suite actuelle ne laisseroit pas d'etre la plus parfaite de toutes les suites possibles, par la raison que Dieu choisit tousjours le meilleur possible." (GP III, 582−583).

the physical universe. The principle of sufficient reason governs the created world; not only does it entail that everything is determinate and intelligible (which for Leibniz means, thinkable), it also entails that everything strives for perfection. Thus the essences that are ideas in the mind of God strive for existence, but only those that constitute this best of all possible worlds succeed; and in the created world, the essences continue to jostle each other, to interfere with each other, as they all strive. This dynamic quality of ideas produces time, as their harmonies produce space; creation entails plurality and mutual limitation, activity and passivity. And the time that is produced is asymmetrical, as creation tends towards greater perfection, a harmonious dissention among the sentient, active individual substances.

What Leibniz heralds is the now received belief that matter is not passive and in-ert, or dead: even a molecule is mobile, active, forceful, and sensitive. As he writes in the *Monadology*, sec. 66–69:

66. (...) there is a world of creatures, of living beings, of animals, of entelechies, of souls in the least part of matter.
67. Each portion of matter can be conceived as a garden full of plants, and as a pond full of fish. But each branch of a plant, each limb of an animal, each drop of its humors, is still another such garden or pond.
68. And although the earth and air lying between the garden plants, or the water lying between the fish of the pond, are neither plant nor fish, they contain yet more of them, though of a subtleness imperceptible to us, most often.
69. Thus there is nothing fallow, sterile, or dead in the universe, no chaos and no confusion except in appearance (...). (GP IV, 618–619)[14]

5 A Thought Experiment

To probe the limits of Leibniz's relationalism, I propose to leave the path of textual analysis for a while, and venture into the forest of thought experiments. Inspired by twentieth century speculation, I propose that we try out Leibnizian relationalism on models of the universe very different from that which he entertained, and see what becomes of the account of time. First, let us suppose that nothing exists except a single particle. Then there is no time, because time is the expression of relations of incompatibility among things and one thing is clearly compatible with itself.

Suppose next that nothing exists except a perfect harmonic oscillator, which moves through a certain series of configurations only to return to exactly the same configuration in which it began. The motion of the harmonic oscillator, with one

---

[14] "66. (...) il y a un Monde de Creatures, de vivans, d'Animaux, d'Entelechies, d'Ames dans la moindre partie de la matiere. 67. Chaque portion de la matiere peut être conçue comme un jardin plein de plantes, et comme un étang plein de poissons. Mais chaque rameau de la plante, chaque membre de l'Animal, chaque goutte de ses humeurs est encor un tel jardin ou un tel étang. 68. Et quoyque la terre et l'air interceptés entre les plantes du jardin, ou l'eau interceptée entre les pois-sons de l'étang, ne soit point plante, ny poisson, ils en contiennent pourtant encor, mais le plus souvent d'une subtilité à nous imperceptible. 69. Ainsi il n'y a rien d'inculte, de sterile, de mort dans l'univers, point de Chaos, point de confusions qu'en apparence (...)" (GP VI, 618–619).

causal state giving rise to another, expresses time, but is the time it expresses finite or infinite? Since its beginning and end state are identical, it seems as if we should identify the times they express; then time would be finite. The local 'befores' and 'afters' would have no global significance; the asymmetry of cause and effect along the way would be absorbed into a larger symmetry, because every effect would ultimately be the cause of the cause… of the cause of its cause. Thus the local in-compatibility of before and after would be absorbed into a global compatibility; but then we must wonder whether this finite time is really temporal at all. It seems that in this picture duration both does and does not occur.

Moreover, the picture seems to contradict the supposition that what exists is a perfect harmonic oscillator, for there is no oscillation. The concept of oscillation involves the notion of repetition, which in turn requires a linear ordering of time, so that when a particular configuration recurs, that is when it occurs again, the first occurrence is earlier than the later one, but the later one is not earlier than the first. We can imagine that the same configuration recurs at a later moment of time; but it is incoherent to suppose that *the selfsame moment of time* recurs at *another moment of time*, for those two moments of time must then be both identified with, and distinguished from, each other. As Leibniz often observes, contradiction makes alleged ideas vanish into nothingness; the relationalist idea of an isolated harmonic oscillator is a mirage, and so is the idea of a moment of time recurring.

So we would have to admit that the time that frames the harmonic oscillator is ongoing, linear and infinite, and so must be constituted by something beyond the relations that hold among the moving parts of the harmonic oscillator; but this goes against Leibnizian relationalism. To avoid this problem, Leibniz must completely fill up his cosmos with things and events that never repeat, on pain of incoherence. Such a cosmos is precisely what his metaphysics provides, chosen by God accord-ing to the Principle of Plenitude, the Principle of Perfection, the Principle of Suffi-cient Reason, and the Principle of Contradiction. Moreover, since all of his monads are body-souls, everything that exists is provided with a developed or rudimentary intentionality, that drives it forward in time. The strong asymmetry observed in the organic, sentient world is guaranteed for everything that exists. In Leibniz's cos-mos, everything is alive and everything strives. The dispute with Clarke shows that Leibniz's cosmos must be a plenum, for otherwise isolated things would show up in absolute space and God's choice of their location would be arbitrary; similarly, if isolated events happened in absolute time, God's choice of when they occurred would be arbitrary. So even if we imagine the ideal harmonic oscillator to express an ongoing, infinite time, perhaps by allowing the natural numbers as a condition of its intelligibility, so that each of its oscillations might thereby be distinguished by a numerical index, it would still violate the Principle of Sufficient Reason.

At this juncture in the argument, however, we might suspect that Leibniz has not discovered the infinity and uni-directionality of time in the relations among things, but merely construed the relations among things so that the time they express will turn out to be appropriate, that is, infinite and uni-directional. And another suspicion may arise: Even if Leibniz is accurately describing the way things are (an organicist, animist plenum), perhaps that in itself sheds no light on time. Time itself may have no flow; and it may prove to be finite, coming to an end that no living thing (including Leibniz) foresees. If our grasp of time is merely empirical, based on temporal relations among things, maybe real time is beyond our grasp. However, for Leibniz no pursuit of truth should be merely empirical; to be a Leibnizian relationalist is not to reduce science to empiricism. Leibniz avoids this skeptical worry by trusting in the ability of metaphysical principles to regulate the interaction of empirical research and theoretical speculation in science. Informing this trust is his trust in the perfection and intelligibility of the cosmos, so that time is the expression of the infinite, harmonious incompatibility of things.

6 Coda

Leibniz understands that productive scientific and mathematical discourse must carry out distinct tasks in tandem: a more abstract search for conditions of intelligibility or solvability, and a more concrete strategy for achieving successful reference, the clear and public indication of what we are talking about. The texts characteristic of successful scientific research will thus be heterogeneous and multivalent. This fact has been missed by philosophers who begin from the point of view of logic, where rationality is often equated with strict discursive homogeneity and method is construed as the rewriting of science and mathematics in a formal, axiomatized language; and it has led scholars influenced by logicism, among them Louis Couturat and Bertrand Russell, to misread Leibniz. While deductive argument is important (since its forms guarantee the transmission of truth from premises to conclusion) as a guide to effective mathematical and scientific reasoning, it does not exhaust method, for Leibniz. As we have seen, Leibnizian method has two dimensions, empirical and rational, and both require analysis, whose logical structure includes abduction and induction, as well as deduction. Moreover, analysis, the search for conditions of intelligibility, is more than logic; it is a compendium of research and problem-solving procedures, which vary among investigations of different kinds of things.

An unswerving focus on logic diverts attention from other forms of rationality and demonstration. Human awareness is both receptive and active, an accommodating construal and an explanatory construction. Some empiricist or naturalist philosophers of science demand that true knowledge be an accurate construal of the way things are, but then they deny the obvious fact that all representation is distortion, however informative it is, and that representation itself changes the way things are. And explanatory analysis goes far 'beyond' the things that invoked it, and thus often sacrifices concrete, descriptive accuracy. Other logicist or anti-realist philosophers of science want to suppose that all knowledge, and indeed all reality, is a human construction, but then they deny the obvious fact that the world is the way it is whether we like it or not, and that it has depths that elude our construals and constructions altogether. Many an explanatory analysis has shipwrecked on the hidden shoals of reality. A more reasonable view of human knowledge is to regard it with Leibniz as a combination of focussed awareness and theoretical elaboration; thus when we combine multiple modes of representation in our scientific work we may in fact have a better chance of doing justice to what we are investigating. Such representational combination and multivocality is just what we find in Leibniz's most important pronouncements on the nature of time.

References

Aiton, E. 1985. *Leibniz. A biography*. Bristol: Adam Hilger.

Belaval, Y. 1960. *Leibniz critique de Descartes*. Paris: Gallimard.

Bertoloni-Meli, D. 1993. *Equivalence and priority. Newton versus Leibniz*. Oxford: Oxford University Press.

Davillé, L. 1909. *Leibniz historien. Essai sur l'activité et la méthode historiques de Leibniz*. Paris: Félix Alcan.

Descartes, R. 1964–1974. In *Oeuvres*, ed. C. Adam and P. Tannery. VIII vol. Paris: Vrin.

Duchesneau, F. 1993. *Leibniz et la méthode de la science*. Paris: Presses Universitaires de France.

Garber, D. 2009. *Leibniz: Body, substance, monad*. New York: Oxford University Press.

Granger, G.-G. 1981. Philosophie et mathématique leibniziennes. *Revue de Métaphysique et de Morale* 86 (1): 1–37.

Grosholz, E. 2011. *"Space and time." The Oxford handbook of philosophy in early modern Europe*, ed. D. Clarke and C. Wilson, 51–70. Oxford: Oxford University Press.

Leibniz, G. W. 1960–1961. *Die Philosophischen Schriften*, ed. C. I. Gerhardt, I–VII vols. Hildesheim: Georg Olms.

Leibniz, G. W. 1966. *Opuscules et fragments inédits*. Hildesheim: Georg Olms.

Newton, I. 1962. *Unpublished scientific papers of Isaac Newton, a selection from the Portsmouth collection in the University Library, Cambridge*, ed. A. R. Hall and M. B. Hall. Cambridge: Cambridge University Press.

Newton, I. 1999. *The PRINCIPIA. Mathematical principles of natural philosophy*. 3rd ed., 1726. Trans: I. B. Cohen and A. Whitman. Berkeley: University of California Press.

Analyticité, équipollence et théorie des courbes chez Leibniz

Eberhard Knobloch

1 Introduction

En 1890 fut créée l'Association Allemande des Mathématiciens. Plus tard, on réalisa pour elle une médaille sur laquelle était écrite l'inscription latine suivante (Fig. 1):

> Artem geometriae discere atque exercere publice interest
> « C'est dans l'intérêt public que d'apprendre et d'exercer l'art de la géométrie. »

Qui y contredirait? Personne assurément. Mais d'où cette l'inscription tire-t-elle son origine? On la retrouve dans le *codex Iustinianus* du droit romain, au chapitre 18 du Livre 9, intitulé: *De maleficis et mathematicis et ceteris similibus* (« Sur les malfaiteurs, les mathématiciens et autres semblables »)! On s'étonne, et à plus forte raison si l'on lit le texte qui suit: « Il est pire de tuer un homme par poison que de le faire mourir par l'épée ». Après quoi se trouve la citation mentionnée plus haut, qui se poursuit ainsi: *Ars autem mathematica damnabilis interdicta est* (« Mais l'art mathématique condamnable est interdit »)!

Pour comprendre cette interdiction, il faut se souvenir que l'art mathématique était alors l'astrologie qui fut interdite par l'empereur romain Dioclétien en 294 (Fögen 1997, p. 13, 260). L'art mathématique et la géométrie sont donc des choses complètement différentes. L'importance de la géométrie, comme le rappelle notre citation, est, elle, incontestée. Mais comment délimiter la géométrie? René Descartes a essayé de donner une réponse à cette question: « je ne sache rien de meilleur que de dire que tous les points de celles [*scil.* les courbes] qu'on peut nommer géométriques, c'est-à-dire qui tombent sous quelque mesure précise et exacte, ont nécessairement quelque rapport à tous les points d'une ligne droite, qui

E. Knobloch (✉)
Technische Universität Berlin, Berlin, Germany
e-mail: eberhard.knobloch@mailbox.tu-berlin.de

© Springer Netherlands 2015
N. B. Goethe et al. (eds.), *G.W. Leibniz, Interrelations between Mathematics and Philosophy,* Archimedes 41, DOI 10.1007/978-94-017-9664-4_5

Fig. 1 La médaille de la Deutsche Mathematiker-Vereinigung (Association Allemande des Mathématiciens). (Eberhard Knobloch, 100 Jahre Mathematik in Berlin. *Mitteilungen der Deutschen Mathematiker-Vereinigung* 2001, p. 34)

peut être exprimé par quelque équation »[1]. La certitude et l'exactitude jouent ici le rôle crucial. L'approche cartésienne concernant l'exactitude était liée à « l'analyse philosophique de l'intuition géométrique » (Bos 2001, p. 411). Nous verrons comment ce point de vue a influencé Leibniz. À cette fin, je voudrais discuter les quatre sujets suivants: le lien entre analyticité et géométrie, l'équipollence, la classification des courbes et la théorie des courbes analytiques.

1.1 Analyticité et géométrie

Notre première question sera la suivante: Que veulent dire « analytique » et « analyticité » chez Leibniz? Quelle est la relation entre ces notions et celle de géométrie? Pour être capable de répondre à cette question, il nous faut étudier l'emploi de l'épithète « analytique » et du substantif « analyse », et cela dans un ordre chronologique. On remarque, en effet, que Leibniz change le sens de sa terminologie avec le temps. De cette manière, nous établirons cinq résultats essentiels.

Nous considèrerons tout d'abord un premier tableau de notions qui sont appelées « analytiques » par Leibniz: *figura analytica* (figure analytique, Janvier 1675)[2], *campus analyticus novae geometriae* (champ analytique de la nouvelle géométrie)[3]; *curva analytica* (courbe analytique, 1675/76)[4], *calculus analyticus exactus* (calcul

---

[1] Texte que Leibniz connaissait par sa traduction latine (Descartes 1659/1661, p. 21): « Aptius quidquam afferre nescio, quam ut dicam, quod puncta omnia illarum, quae geometricae appellari possunt, hoc est, quae sub mensuram aliquam certam et exactam cadunt, necessario ad puncta omnia lineae rectae, certam quandam relationem habeant, quae per aequationem aliquam, omnia puncta respicientem, exprimi possit ».

[2] A VII, 5, 202.

[3] A VII, 5, 193.

[4] Leibniz 1993, 49 (trad. fr. Leibniz 2004, 116 sq.).

analytique exact)[5], *expressio arithmetica sive analytica* (expression arithmétique ou analytique)[6], *methodus certa et analytica* (méthode certaine et analytique)[7], *relatio analytica (vera, generalis)* (relation analytique (vraie, générale))[8]; *aequatio analytica* (équation analytique)[9], *quadratura analytica* (quadrature analytique)[10].

Analysons pas à pas ces notions et d'abord celle de *figura analytica*. Au mois de janvier 1675, Leibniz écrit son étude *De figuris analyticis figurae analyticae quadratricis capacibus* (Sur les figures analytiques donnant lieu à une figure analytique quadratrice). Il y donne la définition suivante: « J'appelle figures analytiques celles dans lesquelles la relation de l'ordonnée à l'abscisse peut être expliquée par une équation » (*Figuras analyticas appello, in quibus relatio ordinatae ad abscissam aequatione explicari potest*)[11]. On remarque que les « figures » sont ici des courbes et non des aires délimitées par des courbes. Leibniz ne dit pas expressément « équation algébrique », mais c'est ce qu'il veut dire. Il se réfère à Descartes en disant qu'il préfère, contrairement à lui, appeler ces figures « analytiques » plutôt que « géométriques ». En d'autres mots, Leibniz remplace la notion de géométrique non pas par la notion d'algébrique (Bos 2001, p. 336), mais par celle d'« analytique ». En fait, sa définition nous rappelle la définition cartésienne d'une courbe géométrique (Descartes 1659/61, p. 21). Descartes avait dit qu'il s'agit des courbes ayant « nécessairement quelque rapport à tous les points d'une ligne droite, qui peut être exprimé par quelque équation » et la traduction latine donnait: *relationem habeant, quae per aequationem aliquam (…) exprimi possit*. Leibniz a donc simplement remplacé *exprimere* par *explicare*.

Dans son traité sur la *Quadrature arithmétique* (1675/1676), Leibniz est plus précis: « J'appelle courbe analytique celle dont tous les points peuvent être trouvés par un calcul exact » (*Curvam analyticam voco cujus puncta omnia calculo exacto possunt inveniri*)[12]. En conséquence, il doit expliquer l'expression « calcul analytique exact ». Le problème est ici simplement déplacé. Dès le début du traité, il est clair que la notion d'algébrique ne coïncide déjà plus avec la notion d'analytique qui est beaucoup plus étendue et qui comprend l'algébricité comme un cas spécial. Leibniz définit: « Mais on appelle ce calcul analytique exact lorsque la quantité cherchée peut être trouvée à partir des données à l'aide d'une équation ayant pour inconnue la quantité cherchée » (*Calculus autem analyticus exactus ille vocatur, cum quantitas quaesita ex datis inveniri potest ope aequationis, in qua ipsa quantitas quaesita incognitae locum obtinet*). Il a remplacé la notion d'équation de la définition du mois de janvier 1675 par la notion de calcul analytique exact. L'équation

---

[5] Leibniz 1993, 50 (trad. fr. 2004, 116 et 117, 12); ainsi que Leibniz 1993, 79 (trad. fr. Leibniz 2004, 218 sq.); Leibniz 1686, 231 (trad. fr. 1989, 138); Leibniz 1714, 394.

[6] Leibniz 1993, 56, 79 (trad. fr. 2004, 138 sq., 216–217, 219).

[7] Leibniz 1993, 107 (trad. fr. 2004, 300 sq.).

[8] Leibniz 1993, 79 (trad. fr. 2004, 217–219).

[9] Leibniz 1682, 119 (trad. fr. 1989, 75).

[10] Leibniz 1682, 119 (trad. fr. 1989, 74).

[11] A VII, 5, 202.

[12] Leibniz 1993, 49 (trad. fr. 2004, 116 sq.).

qui figure dans la définition de cette nouvelle notion peut être une équation infinie ou une série infinie. En d'autres mots, Leibniz généralise la notion d'équation dont le degré peut être fini ou infini et introduit une notion qui est plus générale que la notion d'algébrique et l'englobe. Ce procédé est d'autant plus aisé à effectuer qu'il n'avait pas utilisé expressément l'épithète « algébrique » auparavant. Nous obtenons ainsi un premier résultat essentiel:

1. Analyticité coïncide avec calculabilité.

Ce résultat devient encore plus évident lorsqu'on étudie l'article *De vera proportione circuli ad quadratum circumscriptum in numeris rationalibus expressa* (« Sur la vraie proportion entre un cercle et le carré circonscrit exprimée en nombres rationnels ») qui parut en 1682. Leibniz y remplace le mot *relatio* de son traité sur la *Quadrature arithmétique* par *proportio* et donne une classification des quadratures analytiques (Leibniz 1682, 120; trad. fr. 1989, 75) sous la forme suivante:

Quadrature analytique c'est-à-dire résultant d'un calcul analytique exact		
Transcendante	Algébrique	Arithmétique
$x^x + x = 30$	Au moyen de racines d'équations communes	Exprime la valeur exacte au moyen de séries (infinies)[13]

La notion d'exactitude joue ici à nouveau le rôle crucial. L'exactitude n'est pas réalisée seulement par un calcul exact algébrique, mais aussi par deux autres types d'équations: les équations transcendantes et les équations infinies. En fait, Leibniz distingue entre trois types de quadratures analytiques et ainsi entre trois types d'équations.

Nous reconnaissons un procédé leibnizien d'utilisation d'une notion (par exemple celle d'analyticité ou d'équation) et obtenons le deuxième résultat essentiel:

2. Leibniz conserve une notion mais en change le sens[14].

Après avoir clarifié la signification d'analyticité chez Leibniz nous pouvons élargir le premier tableau des notions qui ont « analytique » comme épithète. Sans vouloir être exhaustif, en voici quelques exemples: *subsidium analyticum*

---

[13] Leibniz 1682, 120:

$$\text{quadratura analytica}$$
$$\text{seu quae per calculum accuratum fit}$$

transcendens	*algebraica*	*arithmetica*
$x^x + x = 30$	*per radices aequationum communium*	*per series (infinitas) exactum exprimit valorem*

[14] La même chose s'applique, par exemple, à la notion d'« indivisible ». Au printemps de l'année 1673, il la définit comme une quantité infiniment petite (A VII, 4, 265) ce qui, au sens strict du mot, porte à contradiction, car d'après la définition aristotélicienne d'une quantité (*Métaphysique* V, 13), ce qui ne peut pas être divisé ne peut pas être une quantité.

(« ressource analytique »)[15]; *valor analyticus* (« valeur analytique »)[16]; *ars analytica* (« art analytique »)[17], *canon analyticus* (« règle analytique »)[18], *res analytica* (« sujet analytique »)[19].

Notre deuxième tableau concerne l'emploi leibnizien du substantif *analysis* (analyse), de nouveau dans un ordre chronologique: *analysis indivisibilium atque infinitorum* (analyse des indivisibles et des infinis)[20]; cette analyse est aussi nommée *calculus differentialis* (calcul différentiel), *calculus indefinite parvorum* (calcul des indéfiniment petits) ou *algebrae supplementum pro transcendentibus* (supplément de l'algèbre pour les transcendantes)[21]. On parvient ainsi un troisième résultat essentiel:

3. L'analyse est un calcul pour Leibniz de même que l'analyticité n'est rien d'autre que la calculabilité.

En conséquence, on trouvera aussi les expressions *analysis infinitorum* (analyse des infinis)[22], *analysis seu ars inveniendi* (analyse ou l'art d'inventer)[23], *analyse des transcendantes*[24]. Dans le titre de (Leibniz 1694a), Leibniz parle d'un nouveau calcul des transcendantes et dans le corps de cet article, il utilise l'expression « analyse des transcendantes » (1694, 308). D'une manière semblable, Leibniz identifie le calcul nouveau avec l'analyse des infinis (Leibniz 1692a, 259) ou parle d'une *analysis infinitesimalium* (analyse des infinitésimaux)[25].

On peut y ajouter une série de termes où la notion d'analyse est caractérisée par une épithète: *analysis tetragonistica* (analyse tétragonistique, 1675)[26]; *analysis transcendens* (analyse transcendante, 1675/76)[27], *analysis pura* (analyse pure)[28], *analysis mea* (mon analyse)[29]; *analysis perfecta* (analyse parfaite)[30]. Leibniz donne la définition suivante[31]: « C'est la marque d'une analyse parfaite lorsqu'un problème peut être résolu ou lorsque son impossibilité peut être démontrée » (*Signum est perfectae analyseos, quando aut solvi problema potest, aut ostendi ejus impossibilitas*). Cette remarque nous rappelle évidemment la fameuse phrase de

---

[15] Leibniz 1684a, 123 (trad. fr. 1989, 88).

[16] Leibniz 1694b, 317 (trad. fr. 1989, 304).

[17] Leibniz 1700, 340 (trad. fr. 1989, 359). Évidemment, l'expression « art analytique » reprend celle de Viète 1591 (page de titre).

[18] Leibniz 1700, 349 (trad. fr. 1989, 382).

[19] Leibniz 1700, 348 (trad. fr. 1989, 379).

[20] Leibniz 1686, 230 (trad. fr. 1989, 137).

[21] Leibniz 1686, 232 s. (trad. fr. 1989, 141).

[22] Leibniz 1689, 242; 1691, 244 (trad. fr. 1989, 192).

[23] Leibniz 1691, 243 (trad. fr. 1989, 192).

[24] Leibniz 1692c, 278 sq.

[25] Leibniz 1713, 412.

[26] A VII, 5, n. 38, 40, 44, 79.

[27] Leibniz 1993, 55 (trad. fr. 2004, 138 sq.); 1703, 362.

[28] Leibniz 1993, 56 (trad. fr. 2004, 138 sq.).

[29] Leibniz 1993, 107 (trad. fr. 2004, 298; 301).

[30] Leibniz 1684a, 123 (trad. fr. 1989, 88).

[31] Ibid.

Viète: « Finalement, l'art analytique s'arroge à bon droit le plus magnifique des problèmes, à savoir de ne laisser aucun problème irrésolu »[32]. La remarque leibnizienne nous révèle en même temps une caractéristique du procédé leibnizien et ainsi un quatrième résultat essentiel:

4. L'idée de perfectionnement ne caractérise pas seulement la philosophie leibnizienne de l'histoire, mais aussi ses mathématiques.

Le panorama commencé plus haut peut être élargi avec les expressions suivantes: *analysis certa et generalis* (analyse certaine et générale)[33]; *analysis interior quaedam* (une certaine analyse plus profonde)[34]; *analysis nova* (analyse nouvelle)[35]. Ainsi se dessine un cinquième résultat essentiel qui nous rappelle les critères cartésiens:

5. Leibniz met en évidence la certitude et la généralité de sa nouvelle analyse.

Il y a encore d'autres adjectifs épithètes qui nous donnent des informations très importantes sur l'analyse et nous permettent de comprendre la classification leibnizienne. Ainsi en 1693, Leibniz déclare: « l'analyse qui correspond à la géométrie des transcendantes (…) est la science de l'infini »[36]; il parle également de « l'analyse ordinaire (…) imparfaite »[37], de « l'analyse ordinaire ou algébrique » (*analysis ordinaria seu algebraica*)[38]; de « l'analyse infinitésimale (…) algébrique » (*analysis infinitesimalis (…) algebraica*)[39]; ou d'un « nouveau genre d'analyse mathématique connu sous le nom de calcul différentiel » (*novum analyseos mathematicae genus, calculi differentialis nomine notum*)[40].

On peut illustrer la classification leibnizienne de la manière suivante:

	Géométrie	
Géométrie de détermination		*Géométrie des mesures Géométrie transcendante*
Algèbre		*Complément de l'algèbre Ressource analytique Supplément de l'algèbre pour les transcendantes*
	Analyse	

---

[32] Viète 1591, 12: « Denique fastuosum problema problematum ars analytice (…) jure sibi adrogat, quod est nullum non problema solvere ».

[33] Leibniz 1686, 230 (trad. fr. 1989, 136).

[34] Leibniz 1689b, 236 (trad. fr. 1989, 165).

[35] Leibniz 1691, 247 (trad. fr. 1989, 199); Leibniz 1692b, 269 (trad. fr. 1989, 220); Leibniz 1714, 395.

[36] Leibniz 1693a, 294 (trad. fr. 1989, 253): « analysis respondens geometriae transcendentium (…) sit scientia infiniti ».

[37] Leibniz 1694a, 307.

[38] Leibniz 1694b, 317 (trad. fr. 1989, 304).

[39] Leibniz 1702, 352 (trad. fr. 1989, 389).

[40] Leibniz 1714, 392.

La géométrie consiste donc en deux parties qui sont toutes deux étudiées à l'aide des deux parties de l'analyse. Ainsi, toute la géométrie est soumise au calcul ou à l'analyse.

La cycloïde peut servir de modèle pour illustrer la manière dont l'attitude leibnizienne envers la relation entre analyticité et géométricité s'est changée au cours du temps. Au mois de janvier 1675, il constate: « Je préfère appeler figures analytiques celle que d'autres après Descartes appellent géométriques. Car je ne vois pas ce qui empêcherait d'appeler la cycloïde géométrique puisqu'elle peut être tracée exactement par un seul mouvement continu et au surplus très simple (…). Mais je conteste qu'elle soit analytique parce que la relation entre les ordonnées et les abscisses ne peut être expliquée par aucune équation »[41]. Cela signifie qu'en 1675 Leibniz justifie pourquoi il remplace l'épithète « géométrique » par « analytique ». D'après le critère cartésien de l'année 1637 pour la géométricité, une courbe est géométrique si elle peut être tracée exactement par un seul mouvement continu. Dans sa *Géométrie*, Descartes admet même plusieurs tels mouvements successifs (Bos 2001, p. 353). En conséquence, la cycloïde devrait être dite géométrique (sans être analytique). A cette époque, l'analyse est encore insuffisante. En plus, il y a encore des courbes qui échappent à la géométricité. L'invention du calcul leibnizien perfectionnera l'analyse et rendra la deuxième constatation fausse. Mais comment peut-on réaliser ou garantir l'exactitude du calcul? Leibniz ne donne pas encore de réponse à cette question.

En fait en 1686, Leibniz répète d'abord cette argumentation: « Il est nécessaire que soient également admises dans la géométrie ces lignes par lesquelles seules ils [*scil.* les problèmes] peuvent être construits. Et parce qu'elles peuvent être tracées exactement par un mouvement continu, comme évidemment la cycloïde et d'autres lignes semblables, il faut les juger en fait (…) géométriques »[42]. Mais la suite du texte montre la différence à l'égard du texte de janvier 1675: « Si y est l'ordonnée d'une cycloïde, $y = \sqrt{2x - xx} + \int dx : \sqrt{2x - xx}$, équation qui exprime parfaitement la relation entre l'ordonnée y et l'abscisse x…. Ainsi, le calcul analytique est étendu à ces lignes qu'on a écartées jusqu'à présent pour aucune autre raison sinon qu'on les en croyait incapables »[43].

L'exactitude joue le rôle crucial dans la détermination de la géométricité. Finalement, les deux notions en viennent à devenir équivalentes chez Leibniz. En 1693, il critique ceux qui mesurent la géométricité seulement au moyen des équations algébriques d'un certain degré: « alors qu'est plutôt géométrique ce qui peut être

---

[41] A VII, 5, 202: « Figuras malim vocare analyticas, quas alii post Cartesium geometricas. Nam cycloeidem exempli gratia non video quid prohibeat appellari geometricam, cum uno continuo motu eoque admodum simplici exacte describi possit (…) analyticam autem esse nego, quoniam relatio inter ordinatas et abscissas nulla aequatione explicari potest ».

[42] Leibniz 1686, 229 (trad. fr. 1989, 134): « Necesse est, eas quoque lineas recipi in geometriam, per quales solas construi possunt (sc. problemata); et cum eae exacte continuo motu describi possint, ut de cycloide et similibus patet, revera censendas esse (…) geometricas ».

[43] Leibniz 1686, 231 (trad. fr. 1989, 138): « Si cycloidis ordinata sit y, fiet $y = \sqrt{2x - xx} + \int dx : \sqrt{2x - xx}$, quae aequatio perfecte exprimit relationem inter ordinatam y et abscissam x…promotusque est hoc modo calculus analyticus ad eas lineas, quae non aliam magis ob causam hactenus exclusae sunt, quam quod ejus incapaces crederentur ».

construit exactement par un mouvement continu » (*cum geometricum potius sit, quicquid motu continuo exacte construi potest*)[44]. En d'autres mots, la construction exacte entraîne la géométricité. Il ne s'agit pas de l'exactitude géométrique: l'exactitude constitue la géométricité. Une année plus tard, Leibniz écrit: « Je constate que tout ce qui est exact, est géométrique, mais mécanique ce qui est effectué par approximation »[45].

En 1695, il défend sa nouvelle définition de l'égalité: Deux quantités sont égales lorsque leur différence est plus petite qu'une quantité quelconque donnée. Il suffit, dit-il, qu'une telle définition soit intelligible et utile pour l'invention, « parce que ce qui peut être trouvé au moyen d'une autre méthode (en apparence) plus rigoureuse, s'ensuit nécessairement toujours d'une manière non moins précise de cette méthode »[46]. Donc, Leibniz met en évidence l'exactitude du résultat au moyen d'une méthode au moins aussi rigoureuse qu'une autre méthode quelconque. En 1714, il résume ainsi sa position: « Au moyen du nouveau calcul, toute la géométrie est désormais soumise dans toute son étendue au calcul analytique » (*Novo calculo jam tota quanta est geometria calculo analytico subjecta est*)[47]. Grâce à son calcul, géométricité et analyticité sont devenues équivalentes.

1.2 Équipollence

Dans son traité sur la *Quadrature arithmétique*, Leibniz introduit la nouvelle relation opératoire d'*équipollence*[48] qui joue un rôle crucial dans son calcul différentiel.

A l'origine, il s'agissait d'une relation entre une ligne droite et une ligne courbe. En fait, il nous faut commencer nos considérations avec Kepler et sa *Nova stereometria doliorum vinariorum* (*Nouvelle stéréométrie des tonneaux à vin*). Il y utilisait cette expression que Leibniz utilisa également dans un premier temps: *aequiparare*. Kepler inscrit un polygone d'un nombre quelconque de côtés dans un cercle. La droite DB est un côté de ce polygone et aussi une corde de l'arc DB. L'arc EB est la moitié de l'arc DB (Fig. 2). Kepler continue: « Mais il est permis de raisonner sur EB comme sur une ligne droite parce que la force de la démonstration coupe le cercle en des arcs minimaux qui sont identifiés (*aequiparantur*) avec des lignes droites »[49].

L'expression *aequiparantur* est cruciale. Kepler ne dit pas « le plus petit arc est égal à une ligne droite », mais il y a une action, une identification de l'arc avec une ligne droite. Pour justifier cette identification, Kepler se réclame de la force de la

---

[44] Leibniz 1693, 290.

[45] Leibniz 1694b, 312 (trad. fr. 1989, 295): « Statuo, quicquid exactum est, geometricum esse, mechanicum vero quod fit approximando ».

[46] Leibniz 1695, 322 (trad. fr. 1989, 328): « cum ea quae alia magis (in speciem) rigorosa methodo inveniri possunt, hac methodo semper non minus accurate prodire sit necesse ».

[47] Leibniz 1714, 394.

[48] Parmentier dans Leibniz 2004, 19.

[49] Kepler 1615, 14: « Licet autem argumentari de EB ut de recta, quia vis demonstrationis secat circulum in arcus minimos, qui aequiparantur rectis ».

Fig. 2 Un polygone inscrit dans un cercle. (Johannes Kepler, *Nova stereometria.* Linz 1615. Dans: Johannes Kepler, *Gesammelte Werke*, Band IX, bearbeitet von Franz Hammer. München 1960, p. 13)

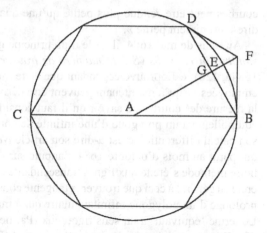

démonstration. Paul Guldin refusa cette référence parce que, disait-il, aucune force de la démonstration ne peut produire un tel fait, en ajoutant à bon droit qu'il n'y a aucun arc minimal; il y a toujours un arc qui est plus petit qu'un segment donné apparemment minimal[50]. La notion de minimal n'était ni bien définie ni définissable. Leibniz a étudié l'œuvre de Guldin *Sur le centre de gravité*: Beaucoup d'allusions dans ses écrits mathématiques l'indiquent[51]. Il se réfère aussi à ce mathématicien suisse au mois de mars ou d'avril de l'année 1673 en étudiant les théorèmes de Guldin sans d'ailleurs le nommer: Il s'y occupe entre autres du centre de gravité d'un hémisphère (A VII, 4, n. 5). Il résume ses considérations en disant: « Mais cette méthode-là est réfutée ainsi: Les courbes doivent y être identifiées avec des polygones à une infinité de côtés »[52].

Leibniz reprend l'expression keplérienne. Environ deux années plus tard, l'idée keplérienne devient la *condition* décisive de ses mathématiques infinitésimales. A l'automne de l'année 1675, il écrit dans sa *Quadrature arithmétique*: « Mais ils (les lecteurs) vont éprouver l'étendue du champ ouvert à l'invention dès qu'ils auront bien compris que toute figure curviligne n'est rien d'autre qu'un polygone comportant une infinité de côtés, de longueurs infiniment petites »[53]. À ce temps-là, Leibniz n'utilisait plus la notion de minimal qu'il avait employée en 1673 mais la notion d'« infiniment petit » bien définie comme « plus petit qu'une quantité quelconque donnée ». L'expression figure curviligne désigne ici une courbe et non pas une aire. On pourrait parler de l'axiome de linéarisation qui n'est pas démontré mais justifié par le théorème 6 du traité: Leibniz y démontre soigneusement que la différence entre certains espaces rectilignes, gradiformes et polygonaux et certaines

---

[50] Guldin 1635–1641 IV, 323.

[51] A VII, 4, 106, 107, 160, 162, 231, 272, 340, 594.

[52] A VII, 4, 63: « Sed ista methodus generaliter ita refutatur: Curvae aequiparandae sunt polygonis laterum infinitorum ».

[53] Leibniz 1993, 69 (trad. fr. 2004, 184–187): « Sentient autem (sc. lectores) quantus inveniendi campus pateat, ubi hoc unum recte perceperint, figuram curvilineam omnem nihil aliud quam polygonum laterum numero infinitorum, magnitudine infinite parvorum esse ».

courbes peut être rendue plus petite qu'une quantité quelconque donnée, c'est-à-dire « infiniment petite ».

Au mois de mai 1684, il parle d'un principe général dans son article *De dimensionibus inveniendis* (*Sur la manière de trouver les mesures (de figures)*, Leibniz 1684a, 126): « J'éprouve cependant que cette méthode et toutes les autres qu'on a employées jusqu'à maintenant peuvent être dérivées de mon principe général pour la mesure des courbes, à savoir qu'il faut considérer une figure curviligne comme équipollente à un polygone d'une infinité de côtés »[54]. La première publication de son calcul différentiel, c'est-à-dire son article *Nova methodus* (*Nouvelle méthode*) qui parut au mois d'octobre 1684 s'appuie sur ce principe: « Il est aussi clair que notre méthode s'étend aux lignes transcendantes... pourvu seulement qu'on s'en tienne en général à ceci que trouver la tangente consiste à tracer une droite... ou le côté prolongé d'un polygone infinitangulaire qui à mes yeux équivaut à une courbe »[55]. Le terme 'équivaut' a un sens rigoureux (Parmentier dans Leibniz 1989, 111 note 59): Deux quantités sont équivalentes si leur différence est infiniment petite.

Jusqu'à maintenant, Leibniz a comparé une courbe avec un polygone infinitangulaire en utilisant les expressions *nihil aliud esse quam* (ne pas être d'autre que), *aequipollere* (être équipollent), *equivalere* (équivaloir). Mais il comparait aussi directement deux courbes en les appelant « équipollentes » sous certaines conditions. Dans son traité *Quadrature arithmétique*, il écrit: « Or j'ai eu la chance de trouver que le théorème énoncé dans cette proposition 7 donne une courbe rationnelle d'une expression très simple et équipollente au cercle; voilà ce qui a donné naissance à la quadrature arithmétique du cercle et à la vraie expression analytique d'un arc à partir de sa tangente pour laquelle nous avons écrit ces choses-là. En conséquence, poussant plus loin mes recherches j'ai trouvé une méthode très générale et belle et cherchée depuis longtemps. À l'aide de cette méthode, une courbe analytique rationnelle peut être exhibée qui est équipollente à une courbe donnée analytique quelconque après avoir réduit le problème à l'analyse pure »[56].

---

[54] « Sentio autem et hanc (methodum) et alias hactenus adhibitas omnes deduci posse ex generali quodam meo dimetiendorum curvilineorum principio, quod figura curvilinea censenda sit aequipollere polygono infinitorum laterum ». Comme dans le traité *Quadrature arithmétique*, Leibniz compare une figure curviligne avec un polygone. Il ne fait aucun sens de comparer une aire avec un polygone. Donc aussi ici, Leibniz désigne par figure curviligne une courbe et non pas une aire comme on a pu le maintenir. (Parmentier dans Leibniz 1989, 111 note 59). En fait, il ne distinguait pas strictement entre *figura* (figure), *linea* (ligne), *curva* (courbe) (voir A VII, 4, Table des matières, articles *curva, figura, Kurve, linea*).

[55] Leibniz 1684b, 223 (trad. fr. 1989, 111): « Patet etiam methodum nostram porrigi ad lineas transcendentes...modo teneatur in genere tangentem invenire esse rectam ducere... seu latus productum polygoni infinitanguli, quod nobis curvae equivalet ».

[56] Leibniz 1993, 56 (trad. fr. 2004, 138 sq.): « Mihi vero feliciter accidit, ut theorema prop. 7. hujus traditum curvam daret rationalem simplicis admodum expressionis; circulo aequipollentem; unde nata est quadratura circuli arithmetica, et vera expressio analytica arcus ex tangente, cujus gratia ista conscripsimus. Inde porro investigans methodum reperi generalem admodum et pulchram ac diu quaesitam, cujus ope datae cuilibet curvae analyticae, exhiberi potest curva analytica rationalis aequipollens, re ad puram analysin reducta ».

La septième proposition est le théorème dit de « transmutation ». Elle décrit comment on peut construire à partir d'une courbe donnée une deuxième courbe de sorte que l'espace délimité par l'axe des abscisses, la seconde courbe et les deux ordonnées extrêmes soit double de celui que délimitent la première courbe et les deux droites reliant ses extrémités au centre de l'angle droit qu'on s'est fixé. En termes modernes, il s'agit d'une intégration partielle. La fin du texte met en évidence l'aspect analytique: *analyse pure* veut dire qu'on n'a besoin ni de l'intuition ni d'une figure. Leibniz n'explique pas expressément la relation d'équipollence entre différentes courbes qui constitue le fondement de la mesure des aires en ce qu'elle peut intervenir entre des figures rectilignes et des figures curvilignes. Mais la citation de la septième proposition ne laisse aucun doute: Deux courbes sont équipollentes lorsqu'elles définissent des aires égales ou lorsqu'elles ont la même quadrature à un facteur multiplicatif près (Parmentier dans Leibniz 1989, p. 20, 39). Dans le cas du théorème de transmutation, c'est le facteur deux. Le cercle $y^2 = 2ax - x^2$ et la *versiera* d'Agnesi $x = \dfrac{2az^2}{a^2 + z^2}$ peuvent servir d'exemple. Les ordonnées de la *versiera* sont commensurables à leurs abscisses (Leibniz à La Roque fin 1675: A III, 1, 346). Pour cette raison, Leibniz l'appelle « analytique rationnelle ». Nous reviendrons sur la classification des courbes dans la section suivante.

Nous constatons donc que l'équipollence des courbes s'appuie sur la notion de quadrature et par conséquent sur la sixième proposition de la *Quadrature arithmétique*. Cette proposition est le théorème fondamental de la théorie leibnizienne de l'intégration (qui est au fond la théorie de l'intégrale riemannienne). Leibniz répète son résultat encore une fois plus tard dans sa *Quadrature arithmétique*: « J'ai donc découvert une méthode par laquelle le cercle de même que toute autre figure peuvent être transformées en une figure rationnelle équipollente et exprimées par des sommes infinies rationnelles... Nous avons reconnu une méthode par laquelle toutes les figures d'équation quelconque peuvent être réduites à des figures rationnelles équipollentes »[57].

Leibniz reprend cette idée cruciale encore une fois dans son article *De geometria recondita et analysi indivisibilium atque infinitorum* (*Sur la géométrie cachée et l'analyse des indivisibles et des infinis*, publié en 1686): « C'est précisément là-dessus que la plupart se sont trompés et se sont barrés la route à ce qui suit, parce qu'ils ne respectaient pas l'universalité propre aux indivisibles de ce type-là comme dx (une progression quelconque des x pouvant être en fait choisie) alors que c'est seulement de cette manière qu'apparaissent d'innombrables transformations et équipollences entre les figures »[58].

---

[57] Leibniz 1993, 110 (trad. fr. 2004, 308, 307, 309): « Mihi ergo methodus innotuit, qua tum circulus tum alia quaelibet figura in aliam rationalem aequipollentem transmutari, ac per summas infinitas rationales exhiberi potest (...). Nos viam deprehendimus qua omnes figurae aequationis cujuscunque reduci possint ad rationales aequipollentes ».

[58] Leibniz 1686, 233 (trad. fr. 1989, 141 sq.): « Nam in hoc ipso peccarunt plerique et sibi viam ad ulteriora praeclusere, quod indivisibilibus istiusmodi, velut dx, universalitatem suam (ut scilicet progressio ipsarum x assumi posset qualiscunque) non reliquerunt, cum tamen ex hoc uno innumerabiles figurarum transfigurationes et aequipollentiae oriantur ».

Nous pouvons résumer: D'abord, Leibniz stipule qu'une courbe peut être identifiée avec un polygone infinitangulaire en utilisant le gérondif *aequiparanda* (A VII, 4, 63), *censenda* (Leibniz 1684a, p. 126). Il ne justifie pas ce principe (Leibniz 1684a, p. 126) dans ses articles publiés mais dans son traité *Quadrature arithmétique* (théorème 6) qu'il a écrit en 1675/76 sans le publier de son vivant. La relation d'équipollence entre deux courbes s'appuie alors sur leurs quadratures; elle est plus générale que l'équipollence d'un polygone et d'une courbe: cette égalité-là est seulement garantie à un facteur multiplicatif près. Leibniz crée même le substantif *aequipollentia* (équipollence) de deux figures. (aires) (Leibniz 1686, p. 233; trad. fr. 1989, 141sq.).

1.3   Classification des courbes

Nous avons vu dans la première partie qu'au mois de janvier 1675 Leibniz remplace la notion de « géométrique » par celle d'« analytique ». Il a fait la même chose avec leurs contraires logiques au mois de décembre 1674 en remplaçant *non geometrica* par *non analytica* (A VII, 5, 139). À cette époque, étant donnée sa première notion d'analytique, il y avait encore pour lui des courbes non-analytiques. En s'appuyant sur cette première notion, il explique: « on peut donc démontrer facilement que le cercle et l'hyperbole n'ont pas de quadratrice analytique parce qu'il est impossible que la quadratrice des anciens et la ligne logarithmique soient analytiques »[59]. C'est au cours de l'année 1675 que sa notion d'analyticité change et qu'il élabore une classification des courbes analytiques en un sens plus large. Dans son étude *De tangentibus et speciatim figurarum simplicium* (« Sur les tangentes et particulièrement des courbes simples ») écrite entre le mois de février et le mois de juin 1676, il définit: « J'appelle une figure *simple* si la nature de cette courbe peut être présentée par une équation de deux termes »[60].

Les exemples qu'il donne montrent que Leibniz classe les courbes analytiques sans le dire expressément. La plus simple de toutes les courbes (analytiques) simples est la ligne droite $y = x$, la deuxième en simplicité est la parabole $ax = y^2$. En général, il s'agit des paraboloïdes $a^z x^v = y^{z+v}$ et des hyperboloïdes $a^{z+v} = x^z y^v$. Dans sa *Quadrature arithmétique*, Leibniz précise: « J'appelle courbe analytique *simple* celle dans laquelle la relation entre les ordonnées et les sections coupées d'un axe quelconque peut être expliquée par une équation de seulement deux termes. »[61] On peut décrire cette famille de courbes par l'équation moderne:

---

[59] A VII, 5, 203: « Hinc facile demonstrari potest circulum et hyperbolam nullam habere quadratricem analyticam; quoniam impossibile est quadratricem veterum, et lineam logarithmicam analyticas esse ».

[60] A VII, 5, 450: « Figuram simplicem voco, cujus curvae natura aequatione duorum terminorum exhiberi potest ».

[61] Leibniz 1993, 51 sq. (trad. fr. 2004, 124 sq.): « Curvam analyticam simplicem voco, in qua relatio inter ordinatas et portiones ex axe aliquo abscissas, aequatione duorum tantum terminorum explicari potest ».

$$v^m = py^n \quad \text{avec } m, n \text{ nombres entiers}$$

Ainsi, Leibniz brise l'unité des coniques, la parabole et l'hyperbole se retrouvant dans une catégorie différente de celle du cercle et de l'ellipse (Parmentier dans Leibniz 2004, 131 note 3).

Leibniz continue cette classification. Dans son étude *De tangentibus et speciatim figurarum simplicium*, il explique: « Une courbe est *rationnelle* lorsqu'un axe quelconque peut être pris tel que, les abscisses et les paramètres étant données en nombres rationnels, les ordonnées perpendiculaires le sont aussi en ces nombres. Et c'est le cas lorsque la valeur de l'ordonnée peut être reçue purement sans aucune extraction de racines (…). Toutes les courbes dans lesquelles les ordonnées sont dans une raison directe ou réciproque multipliée ou multipliée inversement des abscisses sont de cette nature »[62]. Leibniz donne les exemples suivants (A VII, 5, 473 sq.):

$$p^v x^{z-v} = y^z$$

$$x = \frac{ay^2}{a^2 + y^2}, \ y \text{ est l' abscisse}$$

Dans sa *Quadrature arithmétique*, Leibniz reprend: « Une courbe analytique est rationnelle lorsqu'on peut choisir son axe de sorte qu'en partant d'une abscisse et de paramètres rationnels, son ordonnée soit rationnelle »[63]. Les exemples leibniziens sont la parabole et l'hyperbole. Les exemples contraires sont le cercle et l'ellipse. Leibniz mentionne expressément qu'il n'y a dans toute la nature que deux *lignes* possédant des ordonnées rationnelles à la fois selon leurs axes conjugués, à savoir la droite et l'hyperbole ou *figure de l'hyperbole* (Leibniz 1993, 55; trad. fr. 2004, 136 sq.). Cette terminologie démontre de nouveau que Leibniz ne distingue pas strictement entre la notion de ligne, de courbe et de figure.

C'est la raison pour laquelle (en dehors de la ligne droite) la ligne hyperbolique est la plus simple du point de vue de son expression analytique, la ligne circulaire étant la plus simple du point de vue de la construction. Cela veut dire que Leibniz recourt à deux types de simplicité, celui de l'expression (type algébrique) et celui de la construction (type géométrique). Il répète cette classification pour les courbes transcendantes (voir plus bas). Il faut ajouter en effet qu'il y a des courbes simples qui ne sont pas rationnelles comme $y=\sqrt{v}$ et des courbes non-simples qui sont rationnelles comme $y=px^2 + x$. Le cercle et l'ellipse ne sont ni simples (nous le

---

[62] A VII, 5, 473: « Curva rationalis est, cujus directrix aliqua ita assumi potest, ut datis abscissis et parametris in numeris rationalibus, etiam ordinatae normales in numeris haberi possint. Et hoc fit cum valor ordinatae haberi potest pure, sine ulla radicum extractione…Tales sunt omnes curvae, in quibus ordinatae sunt in directa aut reciproca ratione multiplicata aut submultiplicata abscissarum ».

[63] Leibniz 1993, 54 (trad. fr. 2004, 132 sq.): « Curva analytica rationalis est cujus axis ita sumi potest, ut sit ordinata rationalis posito abscissam et parametros, esse rationales ».

savons déjà) ni rationnels parce qu'il est impossible de choisir aucun axe rendant rationnelles les ordonnées. En revanche, les figures rationnelles permettent plus facilement une quadrature exacte ou à défaut arithmétique, c'est-à-dire renfermée dans une série infinie de nombres rationnels (Leibniz 1993, 55; trad. fr. 2004, 138 sq.). Tout hyperboloïde rationnel constitue une quadratrice de l'hyperboloïde de degré supérieur. Tout paraboloïde rationnel constitue une quadratrice du paraboloïde de degré inférieur (Leibniz 1993, 140 sq.; trad. fr. 2004, 192 s.).

Un tableau synoptique peut illustrer la classification leibnizienne:

courbes
géométriques (vs. non-géométriques)
analytiques (vs. non-analytiques)

simples   non-simples
rationnelles   non-rationnelles   rationnelles   non-rationnelles
Par exemple

$$y = pv^2 \qquad\qquad y = \sqrt{v} \qquad\qquad y = px^2 + x \qquad x^2 + y^2 = r^2$$

Il faut souligner que ce schéma ne se trouve pas chez Leibniz. Il s'agit d'un résultat déduit à partir de la terminologie qu'il utilise. Les expressions « non-géométriques », « non-analytiques » sont les compléments logiquement nécessaires des termes techniques « géométriques », « analytiques », mais sont également effectivement employées par Leibniz, comme nous l'avons vu. Le sens des deux paires de termes correspondants change au cours du temps. Avant l'invention de sa nouvelle analyse, Leibniz identifiait géométricité et analyticité. Il y avait pour lui aussi des courbes non-géométriques comme la courbe logarithmique. Après l'invention de sa nouvelle analyse les courbes transcendantes appartiennent à la géométrie. Il n'y a plus de courbes non-géométriques.

Dans ses publications qui parurent à partir de 1682, Leibniz n'utilise plus la notion d'analytique pour classer des courbes. Il introduit la dichotomie algébrique—transcendante (Leibniz 1682, 119; trad. fr. 1989, 75; Leibniz 1684a, 123; trad. fr. 1989, 89; Leibniz 1684b, 223; trad. fr. 1989, 111; Leibniz 1686, 228, 230; trad. fr. 1989, 134, 136 etc.). En fait, Leibniz s'occupait de beaucoup de courbes transcendantes. Je m'en tiendrai à quelques exemples. Au premier rang figurent la cycloïde (Leibniz 1993, prop. 12 et 13) et la courbe logarithmique (Leibniz 1993, déf. après prop. 43, prop. 44, 46, 47, 50). Parmi les courbes transcendantes, la cycloïde peut apparaître comme la plus simple du point de vue de la construction, la courbe logarithmique comme la plus simple du point de vue de l'expression. La première naît du cercle et de l'expression spatiale des angles, la seconde de l'hyperbole et de l'expression spatiale des rapports (Leibniz 1993, 55; trad. fr. 2004, 136–139). En 1684, Leibniz met en évidence les applications immenses de ces deux courbes. Elles peuvent être représentées par des équations de degré indéfini, c'est-à-dire transcendant (Leibniz 1684a, 124; trad. fr. 1989, 90). La cycloïde n'était pas seulement la tautochrone d'Huygens. Elle se révéla aussi être la courbe de la descente la plus rapide, c'est-à-dire la brachistochrone (Parmentier dans Leibniz 1989, 346).

En 1686, Leibniz constate (nous l'avons vu dans la première partie) que la cycloïde peut être tracée rigoureusement par un mouvement continu (Leibniz 1686, 229; trad. fr. 1989, 134) et revient sur ce sujet également plus tard (Leibniz 1693a, 295; trad. fr. 1989, 253 sq.). La courbe logarithmique est un exemple particulièrement intéressant de quadratrice transcendante (dans ce cas de l'hyperbole: Leibniz 1684a, 124; trad. fr. 1989, 90). Pour construire les logarithmes, Leibniz compose un mouvement uniforme et un mouvement retardé par un frottement constant, c'est-à-dire retardé proportionnellement aux espaces parcourus. Il dit lui-même que c'est un moyen physique de construire les logarithmes, alors que la géométrie ordinaire est incapable de les construire exactement (Leibniz 1693a, 295; trad. fr. 1989, 255; Knobloch 2004, 166 sq.).

Autres exemples de courbes transcendantes dont Leibniz s'occupe: les lignes optiques, entre autres la caustique (Leibniz 1689a); l'isochrone (le long de cette courbe un corps pesant tombe uniformément: Leibniz 1689b); l'isochrone paracentrique (sur cette courbe, un corps pesant descendant d'une hauteur H se rapproche ou s'éloigne régulièrement d'un centre A de sorte que les éléments des distances par rapport à A soient proportionnels aux éléments du temps: Leibniz 1694a); la chaînette ou courbe funiculaire (courbe que dessine un fil sous l'effet de son propre poids: Leibniz 1691); la tractrice (courbe telle que la portion de sa tangente comprise entre un point quelconque et l'axe soit constante; Leibniz déclare l'avoir découverte à Paris sur les instances du médecin Perrault (Parmentier dans Leibniz 1989, p. 249); la courbe rhombique ou loxodromique (courbe à double courbure sur une surface sphérique: Leibniz 1691). Leibniz se glorifie d'être capable de calculer selon l'exactitude géométrique l'arc de trajectoire dans un même rhombe, c'est-à-dire de donner la mesure de la courbe rhombique: « C'est une tâche de la géométrie transcendante » (*negotium est geometriae transcendentis*), tandis qu'on ne réalise en général cette mesure que trop peu exactement (*parum accurate*) » (Leibniz 1691, 130 sq.; trad. fr. 1989, 181sq.).

Leibniz avance plus loin: il considère même des courbes tracées au hasard en disant que l'universalité de sa proposition de transmutation est telle qu'elle vaut pour toutes les courbes, même pour les courbes tracées arbitrairement, sans aucune loi déterminée[64]. Dans son *Discours de métaphysique* de l'année 1686, il donne une justification théologique selon laquelle même une telle courbe possède en réalité une équation: « Car supposons par exemple que quelqu'un fasse quantité de points sur le papier à tout hasard (…), je dis qu'il est possible de trouver une ligne géométrique dont la notion soit constante et uniforme suivant une certaine règle, en sorte que cette ligne passe par tous ces points, et dans le même ordre que la main les avoit marqués. Dieu ne fait rien hors de l'ordre et il n'est pas même possible de feindre des événements qui ne soient pas réguliers » (A VI, 4B, 1537 sq.). En d'autres mots, l'ordre du monde créé par Dieu entraîne toujours une certaine topologie. On peut toujours lier deux points l'un à l'autre. Seulement dans un monde qui est mis en ordre moins parfaitement, on ne peut pas lier deux points quelconques.

---

[64] Leibniz 1993, 35 sq. (trad. fr. 2004, 70 sq.).

I. La théorie des courbes analytiques simples

Leibniz distingue les courbes analytiques simples selon une double distinction (Leibniz 1993, 57 sq.; trad. fr. 2004, 144–147): d'un côté les courbes analytiques simples directes (ou paraboloïdes) et les courbes analytiques simples réciproques (ou hyperboloïdes), de l'autre, les courbes analytiques simples rationnelles et les courbes analytiques simples non-rationnelles. Pour l'élaboration de ses trois listes de courbes analytiques simples dans sa *Quadrature arithmétique du cercle*, il s'appuie sur la première dichotomie (Leibniz 1993, 53 sq., 57 sq.; trad. fr. 2004, 130–133, 144–147):

$$p^{m-n} y^n = v^m, \text{ m, n} = 1,2,3 \ etc., \ n < m$$

$$y^n = p^{n-m} v^m, \text{ m, n} = 1,2,3 \text{ etc., m} < \text{n pour les paraboloïdes;}$$

$$y^n v^m = p^{n+m} v^m, \text{ m, n} = 1,2,3 \text{ etc. pour les hyperboloïdes.}$$

Leibniz utilise des puissances du paramètre p pour préserver la loi d'homogénéité de Viète. La permutation des indéterminées y, v et l'abaissement des degrés ou simplification peut faire réapparaître une même courbe sous des équations différentes. Il faut en trouver l'expression la plus simple. Les listes consistent en « classes d'équivalence, dans l'ensemble des équations possibles, modulo les deux opérations de permutation et de simplification » (Parmentier dans Leibniz 2004, 14).

On peut donc illustrer cette classification ainsi:

courbes analytiques simples

paraboloïdes   hyperboloïdes

$y^n = p^{n-m} v^m, n > m$ $y^n v^m = p^{n+m}, \text{m, n} \in \text{IN}$

rationnels non-rationnels   rationnels non rationnels

ax = y $ax^2 = y^3$ xy = a $x^2 y^3 = a$

Après avoir classé ces courbes, Leibniz démontre six théorèmes généraux concernant ces courbes en même temps: théorèmes 15, 16, 17, 18, 21, 22. Avant qu'on puisse les considérer, il faut connaître la notion leibnizienne de figure des sections (*figura resectarum*). Il la définit ainsi (Leibniz 1993, 33; trad. fr. 2004, 64; 67): Par des points quelconques nC d'une courbe $A_1C_2C_3C$ etc., on tire les ordonnées perpendiculairement à l'axe des abscisses. Des points nC on tire les tangentes jusqu'à leurs rencontres avec l'axe des ordonnées. Les points d'intersection sont marqués nT. On transfère les sections AnT (*resectae*) sur les ordonnées nBnC, prolongées si cela est nécessaire, et on obtient les points nD. Les points nD se trouvent sur une nouvelle courbe (quadratrice). L'espace situé entre cette nouvelle courbe, les deux ordonnées et l'axe des abscisses s'appelle « la figure des sections ».

Fig. 3 Une courbe analytique
simple (type 1). (G. W. Leibniz,
*De quadratura arithmetica
circuli* etc., éd. par Eberhard
Knobloch. Göttingen 1993,
p. 49)

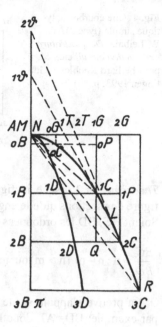

Donc, soit $1C_2C_3C$ une courbe analytique simple $y^n v^m = a$ ou $y^n = v^m b$. Soit $B\theta$ le segment linéaire de l'axe des abscisses entre le point d'intersection de la tangente $C\theta$ et B le pied de l'ordonnée CB. Leibniz distingue entre les trois cas présentés dans les Figs. 3, 4 et 5.

Voici son théorème 15 : $\dfrac{B\theta}{AB}$ (ou sous – tangente : abscisse) $= \dfrac{n}{m}$

C'est le théorème de Ricci (Ricci 1666). Leibniz ne donne aucune démonstration et se contente de dire: « La démonstration demanderait de dépenser beaucoup d'effort » (*Demonstratio multo opus haberet apparatu*)[65], en renvoyant à Ricci. Dans son étude *De tangentibus et speciatim figurarum simplicium* de la première moitié de l'année 1676, il applique ce théorème seulement aux courbes analytiques simples rationnelles: « Sur une courbe analytique simple rationnelle l'intervalle entre la tangente et l'ordonnée prise sur l'axe est à l'abscisse comme l'exposant de la puissance selon laquelle les abscisses sont prises à l'exposant de la puissance selon laquelle les ordonnées sont prises proportionnelles aux abscisses »[66]. Leibniz abandonne la démonstration après quelques pages: il ne donne aucune démonstration complète.

Considérons les cinq autres théorèmes qu'il démontre complètement:

---

[65] Leibniz 1993, 56 (trad. fr. 2004, 140 sq.).

[66] A VII, 5, 474 sq.: « In curva analytica simplice rationali intervallum tangentis ab ordinata sumtum in directrice est ad abscissam ut exponens dignitatis secundum quam sumuntur abscissae ad exponentem dignitatis secundum quam ipsis proportionales sumuntur ordinatae ».

Fig. 4 Une courbe analytique simple (type 2). (G. W. Leibniz, *De quadratura arithmetica circuli* etc., éd. par Eberhard Knobloch. Göttingen 1993, p. 49)

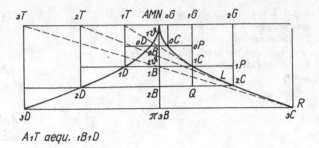

Théorème 16 Lorsque la figure génératrice est une figure analytique simple la figure des sections qu'elle engendrera sera également une figure analytique simple. Soient BC, BD les ordonnées correspondantes des deux figures:

$$\frac{BC}{BD} = n : n - m \ (n > m) \ \text{ou} : m - n \ (n < m) \ \text{si} \ y^n = v^m b; \quad \frac{BC}{BD} = n : n + m \ \text{si} \ y^n v^m = a$$

La preuve s'appuie sur le théorème 15. Pour le cas de la parabole, on obtient par exemple: BD=AT, donc BD : BC=θA : θB, donc $(_1θ_1B - A_1B) : _1θ_1B = _1θ_1B : _1θ_1B - A_1B : _1θ_1B = 1 - m : n = (n - m) : n$

Théorème 17 Le double de $_1CA_2C_1C$ est avec $_1C_1B_2B_2C_1C$ dans le rapport de n−m: n (Fig. 3) ou m−n: n (Fig. 4) ou n+m: n (Fig. 5).

La preuve s'appuie sur le théorème de transmutation et sur le théorème 16. D'après le théorème de transmutation le double de $_1CA_2C_1C$ est égal à la Fig. $_1D_1B_2B_2D_1D$. D'après sa méthode des indivisibles les sommes de toutes les lignes BD et de toutes les lignes BC sont égales aux deux figures considérées dont le rapport est donné par le théorème 16.

Fig. 5 Une courbe analytique simple (type 3). (G. W. Leibniz, *De quadratura arithmetica circuli* etc., éd. par Eberhard Knobloch. Göttingen 1993, p. 50)

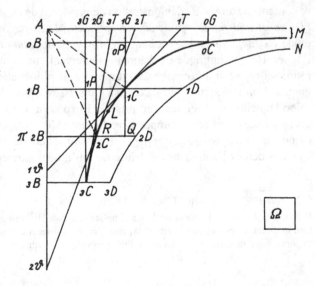

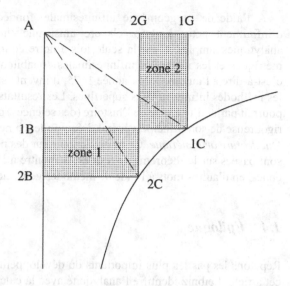

Fig. 6 Les deux zones conjuguées d'une courbe analytique simple. (Eberhard Knobloch, Les courbes analytiques simples chez Leibniz. *Sciences et techniques en perspective* 26 (1993), p. 80)

Théorème 18 La zone entre deux ordonnées, l'arc de la courbe et l'axe est à la zone conjuguée entre les deux abscisses correspondantes, le même arc de la courbe et l'axe conjugué comme n: m (Fig. 6)

La preuve s'appuie sur le théorème 17. Soit S le secteur A_1C_2CA. Donc:

$$2 \text{ fois secteur } S: \text{ zone } 1 = n - m : n \quad \text{ou} \quad m - n : n \quad \text{ou} \quad n + m : n$$

$$2 \text{ fois secteur } S: \text{ zone } 2 = n - m : m \quad \text{ou} \quad m - n : m \quad \text{ou} \quad n + m : m$$

La division de la première équation par la deuxième nous donne le résultat cherché.

Grâce à ces théorèmes, Leibniz peut déduire la quadrature d'une figure analytique simple quelconque en dehors de l'hyperbole conique (théorème 19). Nous mentionnerons les deux théorèmes qui manquent sans les démontrer[67]. Ce qui est intéressant ici est l'approche géométrique qui n'utilise pas encore le calcul différentiel:

Théorème 21 Soit une hyperboloïde donnée. Le rectangle sous l'abscisse infiniment petite A_0B et l'ordonnée infiniment grande $_0B_0C$ est:

  une quantité infinie, si m > n
  une quantité infiniment petite, si m < n,
  une quantité finie, si n = m.

Théorème 22 Soit une hyperboloïde donnée quelconque en dehors de l'hyperbole conique. Il y a deux espaces de longueur infinie $_1C_1BA_0G_0C_1C$ et $_1C_1GA_3B\ldots_3C_1C$. Leurs aires sont infinies ou finies par rapport à l'une des asymptotes selon que n < m ou réciproquement.

---

[67] On trouve une analyse précise des démonstrations dans Knobloch 1993.

À l'aide de sa géométrie infinitésimale, fondée sur la notion bien définie d'infiniment petit, Leibniz élabore ainsi une théorie complète des courbes analytiques simples. C'est la seule fois à notre connaissance: Les méthodes géométriques et les méthodes infinitésimales cohabitent. Presqu'à la même époque, c'est-à-dire à l'automne de l'année 1675, il invente son calcul différentiel qui rend les méthodes infinitésimales superflues. Les résultats obtenus restent valables. On pourrait parler d'une ruse de l'histoire (des sciences) que la justification absolument rigoureuse de son principe général d'équipollence ne se trouve que dans son traité *Quadrature arithmétique*. Ce traité s'appuie sur des méthodes infinitésimales. Elles sont basées sur le théorème six qui est démontré à l'aide de méthodes archimédiennes, en d'autres mots à l'aide de méthodes généralement acceptées.

1.4 Épilogue

Répétons les pas les plus importants du développement intellectuel expliqué dans cet article. Leibniz identifie l'analyticité avec la calculabilité qui devient la notion clé de ses idées à cet égard. Il complète, il perfectionne l'analyse de sorte que la géométrie soit soumise dans toute son étendue au calcul analytique. Cela devient possible grâce à son principe général d'équipollence justifiée rigoureusement dans son traité *Quadrature arithmétique*. Sa classification des courbes reflète ce développement. Finalement il n'y a plus de courbes non-géométriques. Les courbes analytiques simples méritent d'être considérées un peu plus en détail, car Leibniz en a élaboré (une seule fois) la théorie complète.

On pourrait résumer les résultats leibniziens par ses propres mots: « Par l'incursion la plus libre de l'esprit, nous pouvons traiter non moins audacieusement que sûrement les courbes que les droites »[68]. Cette remarque nous rappelle le mot fameux de Cantor: « L'essence des mathématiques consiste justement en leur liberté » (Cantor 1883, p. 182).

References

Bos, Henk. 2001. *Redefining geometrical exactness. Descartes' transformation of the early modern concept of construction*. New York: Springer.
Cantor, Georg. 1883. Über unendliche lineare Punktmannichfaltigkeiten. *Mathematische Annalen* 21:51–58, 545–586. (Cet article appartient à une série d'articles: Mathematischen Annalen 15 (1879), 1–7; 17 (1880), 355-358; 20 (1882), 113-121; 23 (1884), 453–488. Je cite la réédition dans: Georg Cantor, Gesammelte Abhandlungen mathematischen und philosophischen Inhalts, ed. Ernst Zermelo nebst einem Lebenslauf Cantors von Adolf Fraenkel. Berlin 1932, 139–246. (Réédition: Berlin-Heidelberg-New York 1980)).
Descartes, René, and Frans van Schooten, ed. 1659–1661. *Geometria*. Amsterdam: Elzevir.

---

[68] Leibniz 1993, 69 (trad. fr. 2004, 182; 184 s.): « Liberrimo mentis discursu possumus non minus audacter ac tuto curvas quam rectas tractare ».

Fögen, Marie Theres. 1997. *Die Enteignung der Wahrsager, Studien zum kaiserlichen Wissensmonopol in der Spätantike*. Frankfurt a. M.: Suhrkamp.

Guldin, Paul. 1635–1641. *De centro gravitatis*. 4 vols. Vienne: Matthaeus Cosmerovius.

Kepler, Johannes. 1615. Nova stereometria doliorum vinariorum etc. Linz: J. Plank. Je cite la réédition dans Johannes Kepler, Gesammelte Werke, ed. im Auftrag der Deutschen Forschungsgemeinschaft und der Bayerischen Akademie der Wissenschaften. vol. IX. Munich: Beck, 1960, 3–133.

Knobloch, Eberhard. 1993. Les courbes analytiques simples chez Leibniz. *Sciences et techniques en perspective* 26:74–96.

Knobloch, Eberhard. 2004. La configuration (mécanique, géométrie, calcul) et ses bouleversements à la fin du XVIIe siècle. L'exemple de Leibniz. Dans *De Zénon d'Élée à Poincaré, Recueil d'études en hommage à Roshdi Rashed*, eds. Regis Morelon and Ahmad Hasnaoui, 161–174. Louvain: Peeters France.

Leibniz, Gottfried Wilhelm. 1682. De vera proportione circuli ad quadratum circumscriptum in numeris rationalibus expressa. *Acta Eruditorum Février*. 41–46. Je cite la réédition dans: GM V, 118–122. Version française: Leibniz 1989, 61–81.

Leibniz, Gottfried Wilhelm. 1684a. De dimensionibus figurarum inveniendis. *Acta Eruditorum Mai*. 233–236. Je cite la réédition dans: GM V, 123–126. Version française: Leibniz 1989, 82–95.

Leibniz, Gottfried Wilhelm. 1684b. Nova methodus pro maximis et minimis, itemque tangentibus, quae nec fractas nec irrationales quantitates moratur, et singulare pro illis calculi genus. *Acta Eruditorum Octobre*. 467–473. Je cite la réédition dans: GM V, 220–226. Version française: Leibniz 1989, 96–117.

Leibniz, Gottfried Wilhelm. 1686. De geometria recondita et analysi indivisibilium atque infinitorum. *Acta Eruditorum Juillet*. 292–300. Je cite la réédition dans: GM V, 226–233. Version française: Leibniz 1989, 126–143.

Leibniz, Gottfried Wilhelm. 1689a. De lineis opticis et alia. *Acta Eruditorum Janvier* 1689, 36–38. Je cite la réédition dans: GM V, 329–331. Version française: Leibniz 1989, 144–153.

Leibniz, Gottfried Wilhelm. 1689b. De linea isochrona, in qua grave sine acceleratione descendit, et de controversia cum Dn. Abbate de Conti. *Acta Eruditorum Avril* 1689, 195–198. Je cite la réédition dans: GM V, 234–237. Version française: Leibniz 1989, 154–165.

Leibniz, Gottfried Wilhelm. 1691. De linea, in quam flexile se pondere proprio curvat, ejusque usu insigni ad inveniendas quotcunque medias proportionales et logarithmos. *Acta Eruditorum Juin* 1691, 277–281. Je cite la réédition dans: GM V, 243–247. Version française: Leibniz 1989, 186–199.

Leibniz, Gottfried Wilhelm. 1692a. De la chaînette, ou solution d'un problème fameux, proposé par Galilei, pour servir d'essai d'une nouvelle analyse des infinis, avec son usage pour les logarithmes, et une application à l'avancement de la navigation. *Journal des Sçavans Mars*. 1692, 147–153. Je cite la réédition dans: GM V, 258–263.

Leibniz, Gottfried Wilhelm. 1692b. De linea ex lineis numero infinitis ordinatim ductis inter se concurrentibus formata easque omnes tangente, ac de novo in ea re analysis infinitorum usu. *Acta Eruditorum Avril*. 1692, 168–171. Je cite la réédition dans: GM V, 266–269. Version française: Leibniz 1989, 210–221.

Leibniz, Gottfried Wilhelm. 1692c. Nouvelles remarques touchant l'analyse des transcendantes, différentes de celles de la géométrie de M. Descartes. *Journal des Sçavans Juillet*. 1692, 321–322. Je cite la réédition dans: GM V, 278–279.

Leibniz, Gottfried Wilhelm. 1693a. Supplementum geometriae dimensoriae, seu generalissima omnium tetragonismorum effectio per motum: similiterque multiplex constructio lineae ex data tangentium conditione. *Acta Eruditorum Septembre*. 1693, 385–392. Je cite la réédition dans: GM V, 294–301. Version française: Leibniz 1989, 247–267.

Leibniz, Gottfried Wilhelm. 1693b. Excerptum ex epistola G. G. L. cui praecedens meditatio fuit inclusa. *Acta Eruditorum Octobre*. 1693, 476–477. Je cite la réédition dans: GM V, 290–291.

Leibniz, Gottfried Wilhelm. 1694a. Considérations sur la différence qu'il y a entre l'analyse ordinaire et le nouveau calcul des transcendantes. *Journal des Sçavans Août* 1694, 404–406. Je cite la réédition dans: GM V, 306–308.

Leibniz, Gottfried Wilhelm. 1694b. Constructio propria problematis de curva isochrona paracentrica, ubi et generaliora quaedam de natura et calculo differentiali osculorum etc. *Acta Eruditorum Août* 1694, 364–375. Je cite la réédition dans GM V, 309-318. Version française: Leibniz 1989, 282–304.

Leibniz, Gottfried Wilhelm. 1695. Responsio ad nonnullas difficultates a Dn. Bernardo Niewentijt circa methodum differentialem seu infinitesimalem motas. *Acta Eruditorum Juillet.* 1695, 310–316. Je cite la réédition dans GM V, 320–328. Version française: Leibniz 1989, 316–334.

Leibniz, Gottfried Wilhelm. 1700. Responsio ad Dn. Nic. Fatii Duillierii imputationes. Accessit nova artis analyticae promotio specimine indicata, dum designatione per numeros assumtitios loco literarum, algebra ex combinatoria arte lucem capit. *Acta Eruditorum Mai* 1700, 198–208. Je cite la réédition dans: GM V, 340–349. Version française: Leibniz 1989, 359–382.

Leibniz, Gottfried Wilhelm. 1702. Specimen novum analyseos pro scientia infiniti circa summas et quadraturas. *Acta Eruditorum Mai* 1702, 210–219. Je cite la réédition dans: GM V, 350–361. Version française: Leibniz 1989, 383–401.

Leibniz, Gottfried Wilhelm. 1703. Continuatio analyseos quadraturarum rationalium. *Acta Eruditorum Janvier* 1703, 19–26. Je cite la réédition dans: GM V, 361–366. Version française: Leibniz 1989, 402–408.

Leibniz, Gottfried Wilhelm. 1713. (Feuille volante du 29 juillet 1713). Je cite la réédition dans: GM V, 411–413.

Leibniz, Gottfried Wilhelm. 1714. Historia et origo calculi differentialis. GM V, 392–410.

A=Leibniz Gottfried Wilhelm. 1923. Sämtliche Schriften und Briefe, hrsg. von der Berlin-Brandenburgischen Akademie der Wissenschaften und der Akademie der Wissenschaften zu Göttingen. Berlin (depuis 1923 52 volumes en huit séries).

GM=Leibniz, Gottfried Wilhelm. 1962. *Mathematische Schriften*, ed. von Carl Emmanuel Gerhardt. 7 vols. Berlin-London-Halle 1849–1863. (Réédition: Hildesheim 1962).

Leibniz, Gottfried Wilhelm. 1989. La naissance du calcul différentiel, 26 articles des Acta Eruditorum. Introduction, traduction et notes par Marc Parmentier. Préface de Michel Serres. Paris.

Leibniz, Gottfried Wilhelm. 1993. De quadratura arithmetica circuli ellipseos et hyperbola cujus corollarium est trigonometria sine tabulis. Kritisch herausgegeben und kommentiert von Eberhard Knobloch. Göttingen. (Abhandlungen der Akademie der Wissenschaften in Göttingen, Mathematisch-physikalische Klasse 3. Folge Nr. 43).

Leibniz, Gottfried Wilhelm. 2004. Quadrature arithmétique du cercle, de l'ellipse et de l'hyperbole. Introduction, traduction et notes de Marc Parmentier. Texte latin édité par Eberhard Knobloch. Paris.

Ricci, Michelangelo. 1666. *Exercitatio geometrica de maximis et minimis*. Rome: Tinassi.

Viète, François. 1591. In artem analyticen isagoge. Tours. Je cite la réédition dans: François Viète, Opera mathematica, opera atque studio Francisci a Schooten. Leiden 1646, 1–12. (Réédition: Hildesheim 1970).

Leibniz as Reader and Second Inventor: The Cases of Barrow and Mengoli

Siegmund Probst

1 Introduction

During his stay in Paris in the years 1672–1676 Leibniz acquired a wealth of knowledge in mathematics and discovered significant results within a short time. But in respect of some of his findings he had to recognize that he was not the first mathematician to treat them successfully. The best known example is, of course, the calculus, where it was Isaac Newton who anticipated him. But there are other mathematicians who likewise anticipated Leibniz and whose writings were much more easily available to him. During his initial steps towards the calculus in 1673, for example, neither the use of the infinitesimal characteristic triangle, nor the transmutation of curves, nor even recognition of the relationship between the calculation of tangents and areas were completely new insights. Several mathematicians had acquired knowledge of such methods and had worked with them. Indeed, all the examples mentioned had already been published by Isaac Barrow. But even with Leibniz's first mathematical success in Paris, when he solved the problem of the summation of the reciprocal triangular numbers that Christiaan Huygens had set him in 1672, both the specific result and the general method of solution had already been discovered by Pietro Mengoli. Moreover, the general method had also been found by François Regnauld, as Leibniz learned during his visit to London early in 1673. Another example is provided by the arctan series for the circle, which Leibniz formulated in 1673. Unbeknown to him, this series had already been discovered by James Gregory. It was not until April 1675 that he found out about this prior discovery – in a letter from Henry Oldenburg which also contained a sine series of

S. Probst (✉)
Leibniz-Forschungsstelle (Leibniz-Archiv), Hanover, Germany
e-mail: siegmund.probst@gwlb.de

© Springer Netherlands 2015

N. B. Goethe et al. (eds.), *G.W. Leibniz, Interrelations between Mathematics and Philosophy,* Archimedes 41, DOI 10.1007/978-94-017-9664-4_6

Newton[1]. And last but not least, the rules for the quadrature of the higher parabolas and hyperbolas with arbitrary real exponents had been published earlier by John Wallis.

In order to provide data for a comparative study of Leibniz's treatment of predecessors in these topics, it seems necessary first to investigate the extent of Leibniz's knowledge of their results and his use of the sources available to him. This paper aims to contribute to the issue in exploring the cases of two mathematicians who anticipated results found by Leibniz, the more prominent Isaac Barrow (Part I)[2] and the lesser-known Pietro Mengoli (Part II)[3]. Since E. W. v. Tschirnhaus in a letter to Leibniz argued that the calculus provided nothing essentially new in comparison to the methods in Barrows *Lectiones geometricae*[4], the suspicion was raised from time to time that Leibniz had gained benefit in a decisive way from reading this book in finding and developing the differential and integral calculus[5]. After publication of the relevant portion of Leibniz's manuscripts concerning the prehistory and early history of the calculus in the Academy Edition this question can be investigated on a secured basis of original texts[6]. In the case of Pietro Mengoli on the other hand, an investigation of Leibniz's studies on series and on the arithmetic circle quadrature seems especially promising, because Leibniz wanted to publish in this work most of the results mentioned before.[7]

---

[1] Leibniz (1923), A III 1 No. 49$_2$; OC (=Oldenburg 1965) XI No. 2642.

[2] Part I is based on Probst (2011).

[3] Part II is based on a talk "Die Rezeption der Reihenlehre von Pietro Mengoli durch Leibniz in der Zeit seines Parisaufenthalts (1672–1676)" presented at the meeting of the Fachsektion Geschichte der Mathematik der DMV Lambrecht (Pfalz) in 2007 (print forthcoming); an English version entitled "The Reception of Pietro Mengoli's Work on Series by Leibniz (1672–1676)" was presented at the Joint International Meeting UMI-DMV in Perugia (18–22 June 2007).

[4] See Tschirnhaus to Leibniz [April/May 1679] (A III 2 No. 301, 708–712). Barrow (1670), title print in Barrow (1672), title prints with additions Barrow (1674), Archimedes (1675); see Mahnke (1926, pp. 20–22).

[5] The thesis of a dependence of Leibniz's calculus from Barrow was again put forward by J. M. Child in Barrow (1916) and extensively developed in Leibniz (1920). Mahnke (1926), Hofmann (1974, pp. 74–78), and Mahoney (1990, pp. 236–249), denied such a dependence, Feingold (1993, pp. 324–331), repeated Child's claims; Feingold added an investigation of the correspondence and the discussions during the lifetime of Leibniz and of parts of the later research. For a critique see Wahl (2011). The question has been raised again by Blank (2009, pp. 608–609). Recent publications by Nauenberg (2014) and Brown (2012, pp. 58–60), side with Child and Feingold.—A balanced evaluation of the methods and results of Barrow and Leibniz is presented in Breger (2004).

[6] See especially the volumes A VII 4 (1670–1673) and A VII 5 (1674–1676) concerning infinitesimal mathematics.

[7] The studies on series (1672–1676) are printed in A VII 3. The main manuscript text on the arithmetical circle quadrature has been published for the first time completely in Leibniz (1993). Together with the remaining relevant manuscripts from 1673 to 1676, *De quadratura arithmetica* has been published in 2012 in the Academy edition in vol. A VII 6 No. 51, 520–676.

2 Part I: The Reception of Isaac Barrow's *Lectiones Geometricae* (1670) by Leibniz in Paris (1672–1676)

2.1 References to Barrow and Marginal Notes in Leibniz's Copy of the Lectiones Geometricae

Isaac Barrow was one of the first rank of contemporary mathematicians, whose name was known to Leibniz already in his early years in Germany: In *De arte combinatoria* (1666) and in the *Nova methodus discendae docendaeque jurisprudentiae* (1667), he referred to the mathematical symbols that Barrow had used in his edition of Euclid's *Elements* of 1655[8]. In August 1670 Henry Oldenburg informed him of the publication of Barrow's *Lectiones* Opticae (1669) and *Lectiones geometricae* (1670) (A II 1 (2006) No. 27, 99; OC VII No. 1506, 111). Leibniz in a letter to Martin Fogel in January 1671 mentioned only the *Lectiones Opticae* (A II 1 (2006) No. 38, 126–127). Two years later, during his stay in London (January-February 1673) Leibniz acquired the edition of 1672, in which the two works were sold together with a common titlepage (Hanover, Gottfried Wilhelm Leibniz Bibliothek, Leibn. Marg 0)[9]. In his notes on this journey, *Observata in itinere Anglicano*, Leibniz wrote that he had heard that Barrow tackled an optical phenomenon that he had not been able to explain (A VIII 1 No. 1, 6). In April 1673 in a letter to Oldenburg he referred to this statement in Barrow's *Lectiones Opticae* and told him that Huygens and Mariotte declared that they were able to solve the problem concerned (A III 1 No. 17, 87; OC IX No. 2208, 595–596). Another note in the *Observata* could possibly relate to the *Lectiones geometricae* of Barrow, as has been suggested already by Gerhardt[10]; Leibniz wrote: "Tangents to all curves. Development of geometrical figures by the motion of a point in a moving line."[11] Since Leibniz was familiar with the ancient idea of the generation of a line by a flowing point and already in 1671 wanted to construct all possible lines by the composition of rectilinear motions, his note suggests that he was confronted with this issue again in London[12]. The second sentence goes well with a passage on page 27 of the *Lectiones geometricae*, underlined in Leibniz's personal copy: "For every line that lies in a plane can be generated

---

[8] Euclid (1655), „Notarum explicatio", facing page 1; see A VI 1 No. 8, 173; A VI 1 No. 10, 346.

[9] The copy is available online at: http://digitale-sammlungen.gwlb.de/goobit3/ppnresolver /?PPN=688854583. (All pictures of figures in Barrow's *Lectiones geometricae* in this paper are taken from this copy by courtesy of the Gottfried Wilhelm Leibniz Bibliothek Hanover.) The marginal notes to the *Lectiones opticae* are printed in A VIII 1 No. 26, 206–209; the marginal notes to the *Lectiones geometricae* are to be found in A VII 5 No. 43, 301–309; concerning the dating of these notes see 301.

[10] Gerhardt (1891, pp. 157–158); Leibniz (1920, p. 160).

[11] Leibniz (1920, p. 185); "Tangentes omnium figurarum. Figurarum geometricarum explicatio per motum puncti in moto lati." (A VIII 1 No. 1, 5).

[12] The flowing point is already mentioned by Aristotle, *De anima*, 409a 4–5; for Leibniz's discussion of the generation of lines by the composition of rectilinear motions see the *Theoria motus abstracti* (A VI 1 No. 41, 270–271).

by the motion of a straight line parallel to itself, and the motion of a point along it; every surface by the motion of a plane parallel to itself and the motion of a line in it (that is, any line on a curved surface can be generated by rectilinear motions); in the same way solids, which are generated by surfaces, can be made to depend on rectilinear motions."[13] The first sentence of the note in the *Observata* could also refer to Barrow's book where a large part deals with the construction of the tangents of different curves. However, other interpretations are possible: On 8 February 1673, Leibniz took part in a meeting of the Royal Society, during which a letter from René François de Sluse containing an exposition of his method of tangents was read[14]. The letter was published in the current issue of the *Philosophical Transactions*[15], and Leibniz conveyed a copy of the printed version to Huygens in Paris (see A III 1 No. 6, 31–32) and made a personal copy of most of the article. Paraphrasing Sluse's introductory remarks he gave the excerpt the title: "Method to draw tangents to all kinds of curves, without laborious calculation, which can be taught to a boy ignorant of geometry"[16]. The similarities between "tangents to all curves" and "tangents to all kind of curves" are striking. However, motions are not used in Sluse's method of tangents. Another possibility could be a reference to Wallis (1672); this article had been printed a year earlier in the *Philosophical Transactions*. The motion of a point ("motus puncti") is used by Wallis, especially on pages 4014–4016. Perhaps in connection to the reading of Sluse's letter there had been talks where the article by Wallis was mentioned[17].

The rest of the underlined passages in the first part of Leibniz's copy of the *Lectiones geometricae*, which probably originated in the early stages of reading, relates twice (pages 13 and 17) to the concept of motion in geometry, in the third (page 21) Barrow justifies using the terminology of indivisibles (see A VII 5 No. 43, 302). Whether the marginal notes on pages 131–133 and page 136 concerning the classification of curves using their equations already came about at this first reading or only later in Hannover, probably cannot be established. Leibniz uses the equality sign "=" both before mid-1674 as well as from 1677 on. There seems to be no direct evidence for a further reading of the *Lectiones Geometricae* before the autumn of 1675. Only Leibniz's expression of regret in his reply to Oldenburg, dated 12 June 1675, on having heard the news that Barrow had retired from active mathematical

---

[13] Barrow (1916, p. 49); "Omnis, inquam, in uno plano constituta linea procreari potest e motu parallelo rectae lineae, et puncti in ea; omnis superficies e motu parallelo plani, et lineae in eo (lineae scilicet alicujus e rectis modo jam insinuato motibus progenitae) consequenter et linea quaevis etiam in curva superficie designata rectis motibus effici potest" (Barrow 1672, 27; see A VII 5 No. 43, 302). For a comprehensive analysis of Barrow's treatment of curves and motion see Mahoney (1990, 203–213).

[14] Neither Wallis nor Barrow or Newton were present at this meeting, and Leibniz did not meet them during his stay in England or later.

[15] Sluse (1673).

[16] "Methodus ducendi tangentes ad omnis generis curvas, sine calculi laboris, quam etiam puer ἀγεωμέτρητος doceri possit" (A VII 4 No. 6, 70–71).

[17] See the note of the editors to A VII 4 No. 17, 360, which suggests that Leibniz has read this article in spring 1673. It is sure that Leibniz knew the paper in August 1673 (A VII 4 No. 40, 661).

research because of other commitments could be an indication that by now he was familiar with the contents of the book: "I regret that Barrow has done with geometry, for I was still in expectation of many distinguished things from him"[18].

Leibniz probably received new grounds to consider the *Lectiones Geometricae*, when he had several meetings with Tschirnhaus in October 1675. His compatriot, who had recently arrived from England[19], owned a copy of the edition of *Lectiones geometricae* with additions printed on pages 149–151[20]. Leibniz noted at the end of his copy of the 1672 edition, in which these additions are missing, that he had seen these "addenda"[21]. This note as well as the marginal notes on page 85 and the note on the related figures No. 122 and 125 in Leibniz's personal copy may have originated in the context of these meetings. The single marginal note to the text of page 85 says, "I know for some time" ("Novi dudum")[22], and on the pages with figures No. 122 and 125, corresponding to the text on pages 85–89, Leibniz expressed some of the results of Barrow with his new integral symbol. The use of the integral symbol shows that these notes were not written earlier than the end of October 1675 (Fig. 1).[23]

2.2 Readings Without References to Barrow

Leibniz read at least selectively Barrow's *Lectiones geometricae* during the following months. This is documented by his marginal notes and additions to figure No. 119 concerning the quadratrix curve (Fig. 2).

The marginal notes (two equations) were written first and are partly overwritten by the additions to the figure which are made in a different ink. The notes are probably related to a manuscript from June 1676, *De Quadratrice* (A VII 5 No. 86), which is based on the investigation of the quadratrix by Barrow: Leibniz sketches a similar figure in his manuscript: several points are designated with the same letters and he adopts two equations directly from the text of Barrow. His additions to figure No. 119 in the *Lectiones geometricae* are, however, related to a manuscript written

---

[18] OC XI No. 2672, 333; "Barrovium geometrica missa fecisse doleo; nam multa ab eo praeclara adhuc exspectabam." (A III 1 No. 55, 256.)

[19] See Mayer (2006).

[20] See J. Collins to J. Gregory, 19/29 October 1675, printed in: Turnbull (1939, p. 342). Perhaps Tschirnhaus owned the edition of the *Lectiones geometricae* that had been added to Barrow's edition of Archimedes (1675). There are notes from a talk between Leibniz und Tschirnhaus in February 1676 that refer to this edition (A VII 1 No. 23, 180–181).

[21] The additions contain solutions to three problems concerning the arc length of curves, a generalization of the quadrature of the cycloid and several propositions on maxima and minima based on tangent properties.

[22] Barrow's theorem XI, I on the area under the curve of the subnormals corresponds to $\int yy'dx = \int ydy = \frac{y^2}{2}$. Leibniz proved an equivalent proposition in 1673 (A VII 4 No. 27, prop. 6, 467–468).

[23] The notes on the right side of Fig. 123 refer to Fig. 125.

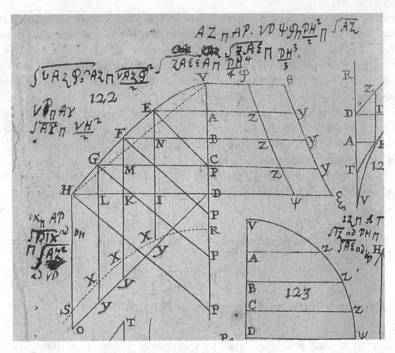

Fig. 1 Leibniz's marginal notes to Barrow's Fig. 122

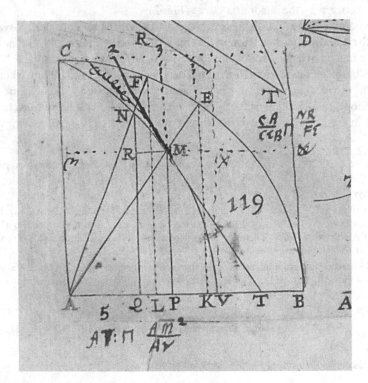

Fig. 2 Leibniz's notes and additions to Barrow's Fig. 119

in Hannover, *De Quadratura quadratricis* (LH 35 XIII 1 fol. 236–239), dating from 6/[16] July 1677.

Another example from spring of 1676 is the *Praefatio opusculi de Quadratura Circuli Arithmetica* (A VII 6 No. 19, 169–177; GM V, 93–98): This is the only manuscript of the Paris period known so far where Leibniz mentions the circle approximation published by Adriaan Metius (A VII 6 No. 19, 173; GM V, 95)[24]. In a small note in this manuscript, not included in Gerhardt's edition, Leibniz wrote down two inequalities. These two inequalities are the results of the approximation method by which Barrow derived the result of Metius in the *Lectiones geometricae*. Barrow expressed the results (for a circle with a diameter of 113 units) in the following words: "the whole circumference, calculated by this formula, will prove to be greater than 355 less a fraction of unity", and "the whole circumference is less than 355 plus a fraction"[25], Leibniz used symbolic formulas: "$c \ulcorner 355 - \dfrac{1}{b}$" and "$c \urcorner 355 + \dfrac{1}{b}$"[26].

These two instances where it is sure that Leibniz used Barrow's book without reference are from the year 1676 and therefore could not have any influence on Leibniz's invention of the calculus which took place earlier. In addition, their thematic relevance for the calculus is a minor one. But the fact that Leibniz does not refer to his source in both cases, suggests further investigation into his manuscripts of 1673–1675. Perhaps there can be more adoptions from Barrow than these two. We know that in the use of the infinitesimal characteristic triangle, the transmutation of curves, and the insight into the relationship between the determination of tangents and of areas Barrow had preceded Leibniz[27]. If there is any adoption of Barrow's methods and results concerning the invention of the calculus it should be possible to discover it in manuscripts dealing with these topics.

2.3 Transmutation Method and Characteristic Triangle

With regard to the transmutation method and the characteristic triangle of Leibniz, whose dependence on Barrow J.M. Child had claimed[28], D. Mahnke has defended the independence of the development of Leibniz[29]. His argument on the basis of then unpublished manuscripts can now be confirmed by means of those texts recently published in A VII 4[30].

---

[24] The circle approximation $\pi \approx \dfrac{355}{113}$ found by Adriaan Anthonisz (ca. 1543–1620) was published by his son Adriaan Metius in Metius (1625, 178–179), and in Metius (1633, 102–103).

[25] Barrow (1916, 150 and 151); "tota circumferentia major quam 355, minus fractione unitatis" and "fore Totam circumferentiam minorem quam 355, plus fractione" (Barrow 1672, 103).

[26] A VII 6 No. 19, 172; the symbols $\ulcorner$ and $\urcorner$ are equivalent to the modern symbols > and < for "greater than" and "smaller than".

[27] For Barrow's transformation methods see Mahoney (1990, 223–235).

[28] See Leibniz (1920), especially 15–16, 172–179.

[29] Mahnke (1926, 8–43).

[30] See xxii–xxiii in the introduction to A VII 4.

The transmutation method of Leibniz is a special method of integral transformation and proceeds in two steps: first, the decomposition of a curve segment into infinitesimal triangles starting from a common endpoint and infinitesimal baselines, the elements of arc, which together form the arc of the curve segment. Second, the use of the similarity of the infinitesimal triangle consisting of the elements of abscissa, ordinate, and arc (or tangent) that Leibniz calls the characteristic triangle of the curve and certain finite triangles (e.g., the triangle formed by the ordinate, tangent and subtangent of the curve), which allows the establishing of proportional equations between finite and infinitesimal sides of the triangles considered. Based on his investigation of surfaces of revolution and the associated determination of arc lengths Leibniz pursues in the spring of 1673 the idea to divide the area under a curve into triangles with a common vertex in the center of gravity of the arc and infinitesimal bases on the arc of the curve (A VII 4 No. 5, 63–64). After that he tries several approaches to implement this idea using the example of the parabola and the circle (A VII 4 No. 5, 64–69; A VII 3 No. 17, 202–227; A VII 4 No. 10_1, 140; A VII 4 No. 10_2, 156–158; A VII 4 No. 12_1, 174–176). Later Leibniz learns about the results of the rectification method of H. van Heuraet (A VII 3 No. 16)[31] and immediately tries to form infinite series of numbers whose sum would, for example, give a result for the rectification of an arc of a parabola (A VII 3 No. 17).

Leibniz uses different starting points for the decomposition of the area, from centers of gravity he moves to any point on the axis of the curve and finally uses the apex of the curve, drawing chords from the apex to the points on the arc of the curve as Barrow had done before him (*Lectiones geometricae* XI, § XXIV, 92). This means that only after a series of general considerations and several investigations did Leibniz arrive at the point where Barrow started his transmutation. Since the area of an infinitesimal triangle with an element of arc or tangent as a baseline and the vertex at the apex of the curve is determined, when the altitude of this triangle (i.e. the perpendicular from the apex to the tangent) is determined, Leibniz gains from this the following segment theorem: The area of the curve segment is equal to the sum of the areas of these triangles and therefore equal to half the sum of the products from the baselines and altitudes of these triangles. It happened a few times that Leibniz forgot to halve the sum of the infinitesimal rectangles (e.g., A VII 4 No. 16, 271). Using perpendiculars from the axis to the tangent Leibniz forms different right triangles between axis, tangent and perpendiculars to the axis or to the tangent (e.g., A VII 4 No. 5, 64; A VII 4 No. 10_2, 156; A VII 3 No. 17, 202–203, 210, 220, 222; see Fig. 3). By contrast, Barrow formed his right-angled triangles with perpendiculars to the chords (Fig. 4):

---

[31] Leibniz marks neighbouring points on the abscissa in the related drawing and calls the distances between these points arbitrarily small ("quantumvis parvae"), but in the following he only deals with the ordinates (A VII 3 No. 16, 200). He does not seem to have noticed the characteristic triangle used by van Heuraet. Perhaps Leibniz at this time did not yet know the original publication Heuraet (1659), but used the presentation of the result in Huygens (1673, 69–73). The related drawings in this book (p. 70–71) do not contain characteristic triangles, and Leibniz added them on page 70 in his own copy of the book he had received from Huygens (A VII 4 No. 2, 32).

Fig. 3 A VII 3 No. 17₃, 220

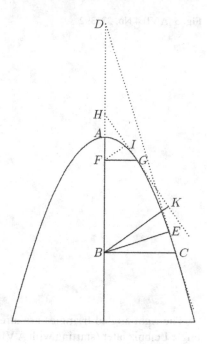

Fig. 4 Barrow's Fig. 129

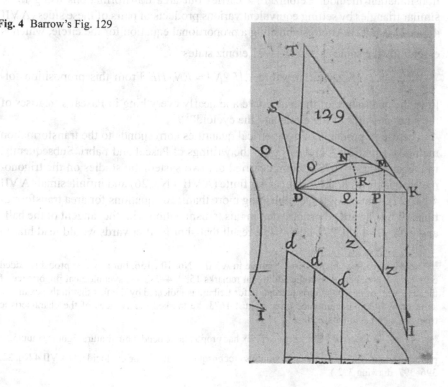

Fig. 5 A VII 4 No. 22, 392

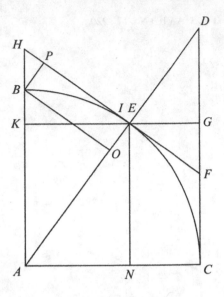

The right-angled triangles are already similar triangles to the infinitesimal triangle Leibniz later (starting with A VII 4 No. 24[32]) calls the "characteristic" triangle of the abscissa, ordinate and tangent differences, and which is the basis for his transmutation method. Leibniz first carries out area transformations using finite similar triangles by setting equivalent various products of pairs of these sides (A VII 4 No. 21 and 22). After establishing a proportional equation for the circle, which is essentially the same as $\tan = \dfrac{\sin}{\cos}$, Leibniz states:

" $\dfrac{AE}{EN} = \dfrac{HE}{NA}$ [Fig. 5]. Therefore, $AE \cdot NA = EN \cdot HE$. From this proposition follows the quadrature of the sine curve and nearly everything in Pascal's treatises of the sines and arcs of the circle and the cycloid"[33].

The use of products of geometrical quantities corresponds to the transformation methods, which he found earlier in the writings of Pascal and Fabri. Subsequently, in the summer of 1673, Leibniz carried out two systematic studies on the trigonometric quantities in the circle, using finite (A VII 4 No. 26) and infinitesimal (A VII 4 No. 27) right triangles, establishing more than 150 equations for area transformations. While he already succeeded in this transmutation with the tangent of the half-angle (A VII 4 No. 27, 489–494) a result that shortly afterwards would lead him to

---

[32] There is already a characteristic triangle in A VII 4 No. 10_2, 156, but this was probably added only later in connection with the additional remarks 158 l. 4–7. The psychological importance of the discovery of the characteristic triangle for Leibniz is indicated by the fact that in this example, as in many others documented after August 1673, he marked the vertices of the characteristic triangles with his initials *G, W, L*.

[33] " $\dfrac{AE}{EN} = \dfrac{HE}{NA}$ Ergo $AE \cap NA = EN \cap HE$. Ex hac propositione pendet quadratura figurae sinuum, et pleraque omnia in Pascalii tract. de sinubus arcubusque circuli, deque cycloeide." (A VII 4 No. 22, 396–397; drawing 392.).

the circle series (A VII 4 No. 42), he formulated transmutation theorems for general curves also with other quantities (A VII 4 No. 27, 495; A VII 4 No. 39_1, 617 and 621). Leibniz uses the term "characteristic triangle" for the first time not in the case of the circle, but rather with other curves such as the conchoid curves, ellipses and cycloids (A VII 4 No. 24, No. 28, No. 29). Perhaps Leibniz coined the term after he noted a theorem of J. De Witt[34] as "ellipsis character" (A VII 4 No. 28, 502).

2.4   Barrow's Prop. XI, § XIX, and Leibniz's Theorem

In the *Lectiones geometricae*, Barrow demonstrates some rules, which are equivalent to rules of differential calculus: for example, VIII, § IX (quotient rule); IX, § XII (product rule)[35]. However, the texts in which Leibniz himself derives these rules (A VII 5 No. 51_2, No. 70, No. 89), do not seem to depend immediately on Barrow's publication. Although Barrow's text contains equivalents to the quotient rule and the product rule, the two rules are not formulated explicitly. The former theorem is a rule for determining geometrically the tangent of a curve if the tangent of the curve with reciprocal ordinates is known. The second is embedded in a more general theorem on the construction of tangents of curves that form the geometric means.

When Leibniz records the first example of a simple case of the product rule of differentiation in his new notation on 27 November 1675, he calculates with infinitesimal differences and remarks: "Now this is a really noteworthy theorem and a general one for all curves. But nothing new can be deduced from it, because we had already obtained it."[36] Apparently he immediately realized that the statement is equivalent to another one he had used since the spring of 1673, expressing the relations of the area of a segment of a curve and its complement to the circumscribed rectangle[37], in modern notation $xy = \int y\,dx + \int x\,dy$. It is noteworthy that Leibniz examines in this text the relationship between integration and inverse tangent method, and even in the same paragraph carries out transformations that are based on the equality of the ratio of the infinitesimal quantities dx and dy with the ratio of the subtangent t and the ordinate y. But he obtains the result directly from the investigation of sums (integrals) and differences of abscissas and ordinates of curves.

The situation is similar with his derivation of the quotient rule: In the spring of 1673 Leibniz calculates the differences of the terms of the sequence $\dfrac{1}{a^2}$ (A VII 3 No. 13, 160), in August 1673 the differences of the ordinates of the hyperbola $y = \dfrac{a^2}{c+x}$ (A VII 4 No. 40_2, 683). Again, the quotient rule is the result of a calcula-

---

[34] Witt (1659), especially book I, prop. 18, 224.

[35] See Breger (2004, 199–200).

[36] Leibniz (1920, 107); "Quod Theorema sane memorabile curvis omnibus est commune. Sed nihil novi ex eo ducetur, quia adhibuimus jam." (A VII 5 No. 51_2, 365.).

[37] See A VII 4 No. 10, 136; A VII 4 No. 40, 690 and 705; more detailed in A VII 3 No. 40, 578–579 (October 1674—January 1675).

Fig. 6 Barrow's Fig. 127

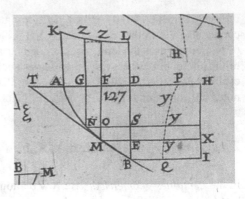

tion of differences: Leibniz subtracts fractions that express neighbouring ordinates (A VII 5 No. 70, 506; A VII 5 No. 89, 593–595). The consideration of tangents is irrelevant to this.

But there is a topic based on the consideration of tangents that is of interest for the invention of differential and integral calculus: the discovery of the equivalence of solutions for the inverse tangent problem and the problem of the determination of the area under a curve. This is what is often called the geometric form of the fundamental theorem of differential and integral calculus. N. Guicciardini, in a review of the volumes A VII 4, and VII 5, recently called attention to a possible influence of Barrow on Leibniz in regard to this issue[38].

Barrow, who was the first to publish such a theorem, put the two reciprocal statements into two separate theorems (*Lectiones geometricae* X, § XI, 78 and XI, § XIX, 90–91). The second is illustrated by his Fig. 127 (Fig. 6):

"Again, let AMB be a curve of which the axis is AD and let BD be perpendicular to AD; also let KZL be another line such that, when any point M is taken in the curve AB, and through it are drawn MT a tangent to the curve AB, and MFZ parallel to DB, cutting KZ in Z and AD in F, and R is a line of given length, TF: FM=R: FZ. Then the space ADLK is equal to the rectangle contained by R and DB."[39]

The subtangent *TF* of *AMB* is to the ordinate *FM* of *AMB* as a constant *R* to the ordinate *FZ* of *KZL*. So we have $FZ=R \times FM/TF$. If $FM=y$, then results $FT=y/y'$,

---

[38] Guicciardini (2010, p. 546): "The marginalia to Isaac Barrow's *Lectiones geometricae* (1670) are particularly noteworthy, since Barrow's lectures would have provided Leibniz with a geometric expression of the so-called fundamental theorem of the calculus." Nauenberg (2014) argues for an influence of Barrow's *Lectiones geometricae* on Leibniz in the case of this theorem on the basis of an analysis of Leibniz (1693). Unfortunately Nauenberg does not investigate any of the manuscripts from 1673 (published in A VII 4) nor does he notice occurrences of similar statements in the manuscripts from 1674 to 1676 (e.g. A VII 5 No. 26, 203–204, and A VII 5 No. 49, 348).

[39] Barrow (1916), 135; "Porro, sit curva quaepiam AMB, cujus axis AD, & huic perpendicularis BD; tum alia sit linea KZL talis, ut sumpto in curva AB utcunque puncto M; & per hoc ductis rectâ MT curvam AB tangente, rectâ MFZ ad DB parallelâ (quae lineam KL secet in Z, rectam AD in F) datâque quâdam lineâ R; sit TF . FM :: R . FZ; erit spatium ADLK aequale rectangulo ex R, & DB."

Fig. 7 A VII 4 No. 40₃, 692

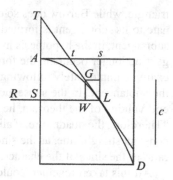

and thus $FM/TF=y'$ and $FZ=R \times y'$. The curve AMB is therefore an antiderivative of the curve KZL. Barrow proved this as follows:

"For, if DH=R and the rectangle BDHI is completed, and MN is taken to be an indefinitely small arc of the curve AB, and MEX, NOS are drawn parallel to AD; then we have NO : MO = TF : FM = R : FZ; NO . FZ = MO . R, and FG . FZ=ES . EX. Hence, since the sum of such rectangles as FG . FZ differs only in the least degree from the space ADLK, and the rectangles ES . EX form the rectangle DHIB, the theorem is quite obvious."[40]

Leibniz formulates an equivalent theorem in August 1673:

"Let LD be a curve [Fig. 7], its sine (i. e. the ordinate perpendicular [to the abscissa AS]) SL, abscissa AS, tangent TL, characteristic triangle GWL. And if it happens that ST is to SL or GW is to WL as a certain constant straight line [c is to] the corresponding sine SR of another curve with the same axis (i. e. the same abscissa), then any portion of the area below the other curve, cut off by its sine, will be equal to a rectangle formed by SL in c. The demonstration of this is very easy: $\frac{TS}{SL} = \frac{GW}{WL}$ or $\frac{g}{w}$ from construction; $= \frac{c}{RS=r}$ from presupposition, therefore $cw=rg$. This suffices as proof for those understanding these matters."[41]

The theorems of Barrow and Leibniz differ mainly by the fact that Leibniz already in the formulation of the statement introduces the infinitesimal characteristic

---

[40] Barrow (1916, p. 135); "Nam sit DH=R; & compleatur rectangulum BDHI; tum assumptâ MN indefinite parvâ curvae AB particulâ ducantur NG ad BD; & MEX, NOS ad AD parallelae. Estque NO . MO :: TF . FM :: R . FZ. Unde NO×FZ=MO×R; hoc est FG×FZ=ES×EX. ergò cum omnia rectangula FG×FZ minimè differant à spatio ADLK; & omnia totidem rectangula ES×EX componant rectangulum DHIB, satis liquet Propositum."

[41] "Sit curva LD, cuius sinus (id est ordinata normalis) SL, abscissa AS, tangens TL, triangulum characteristicum GWL. Sique fiat ut ST ad SL, vel GW ad WL, ita recta quaedam constans [c ad] alterius cuiusdam figurae eiusdem axis sinum respondentem (respondentem inquam[,] id est eiusdem abscissae) SR, portio quaelibet ab altera figura abscissa per sinum eius, aequabitur rectangulo SL in c. Cuius rei demonstratio haec est perfacilis: $\frac{TS}{SL} = \frac{GW}{WL}$ vel $\frac{g}{w}$ per constructionem; $= \frac{c}{RS=r}$ ex hypothesi, ergo $cw=rg$. Quod rerum harum intelligentibus sufficit ad demonstrationem." (*Pars III^{tia} Methodi tangentium inversae et de functionibus*, A VII 4 No. 40₃, 692–693.) – Neither Gerhardt (1848, 20–22), or Gerhardt (1855, 55–57), nor Mahnke (1926, 43–59), mention this theorem in their investigations of the manuscript.

triangle, while Barrow does so only in the proof. In both cases the ratio of the ordinate to its subtangent is formed without using a terminological designation for the subtangent; in the hypothesis in each case only the tangent is employed, but the tangent no longer occurs in the formulation of the proportional equations. Only in the example immediately following does Leibniz employ his usual term "producta" for the subtangent. In the subsequent conclusion, with which he emphasizes the general validity of the theorem, he writes by mistake "tangent" instead of "producta": "Therefore, the quadrature of all curves can be obtained, whose sines are to a certain constant straight line, as the sine of another known curve is to its tangent; or as the ratio of the sines of the characteristic triangle of a known curve".[42]

All this taken together, could create the impression that Leibniz formulated his theorem on the model of Barrow's [43]. This possibility cannot be completely ruled out, but it is to be noted that the statement $\frac{TS}{SL} = \frac{GW}{WL}$, which is required for the formulation of the theorem, is based on the similarity of the characteristic triangle with the triangle of subtangent, ordinate and tangent, and is already pronounced by Leibniz in the beginning of the manuscript (A VII 4 No. 40_1, 657). He expresses the relationship between ordinate and subtangent in the following way, letting the difference of the abscissas be equal to the unit: "In short, the matter goes back to the following: The straight line ED, the ordinate, divided by ID, its difference to the preceding ordinate, gives the straight line ME [scil: the subtangent]" [44]. Leibniz then refers to this relation in A VII 4 No. 40_3, 689, shortly before the formulation of the theorem. The second half of the theorem $\frac{GW = g}{WL = w} = \frac{c}{RS = r}$ does not follow immediately from the preceding considerations. A similar proportional equation had been

obtained by Leibniz in his previous investigations of the characteristic triangle of the circle, the radius playing the role of the constant term c (A VII 4 No. 27, prop. 3, 467), but he had not attempted a generalization there. He studies the same example among others again in A VII 4 No. 40_3, 696–697, but his calculations probably were only carried out after the formulation of the theorem. It is therefore likely to remain an open question whether Leibniz found this theorem in the course of his studies on the inverse tangent method entirely independent from Barrow, or whether he had encountered the theorem while reading Barrow's book.

Overall, it should be noted that Leibniz counted Barrow among the great mathematicians in the prehistory of calculus, as is evident by his statement from 1 November 1675 in *Analyseos tetragonisticae pars tertia*: "Most of the theorems of the geometry of indivisibles which are to be found in the works of Cavalieri, Vincent,

---

[42] "Quare omnium figurarum haberi potest quadratura, quarum sinus sunt ad rectam quandam constantem, ut sunt sinus alterius cuiusdam figurae cognitae ad suam tangentem; seu ratio sinuum trianguli characteristici figurae cognitae."

[43] It should be taken into consideration that both theorems are formulated in analogy to the rectification theorem of Heuraet (1659, 518): Heuraet used the normal of the curve for the proportional equation, Barrow and Leibniz use the subtangent.

[44] "Breviter res eo redit: Recta *ED*, applicata, divisa per *ID*, differentiam ab ipsamet et applicata praecedente, dat rectam *ME*." (A VII 4 No. 40_1, 660); see also Mahnke (1926, 44–46).

Wallis, Gregory and Barrow are immediately evident from the calculus"[45]. Beyond this, Leibniz doesn't seem to have acknowledged any influence of Barrow's writings on his discovery of the calculus, neither then nor later. In 1686 he places Barrow alongside James Gregory in *De geometria recondita* when he writes: "These were followed by the Scotsman James Gregory and the Englishman Isaac Barrow, who in famous theorems of this kind advanced science in a wonderful way."[46]

3   Part II: The Reception of Pietro Mengoli's Work on Series by Leibniz (1672–1676)

Probably in September 1672 Leibniz, in a discussion with Christiaan Huygens, expressed his opinion that he possessed a general method for finding the sum of infinite series. The basis for this was his realization that the terms of a monotonously decreasing zero-sequence are equal to the sums of the differences of the following terms. Huygens tested the mathematical abilities of Leibniz by proposing to him that he find the sum of the series of reciprocal triangular numbers, a result he had found himself some years before, in 1665, but had not published[47].

After a few futile attempts, Leibniz, within a short space of time, achieved success: he found out that the reciprocal triangular numbers are the doubled differences of the harmonic series (A VII 3 No. 1 and 2). In addition he was able to calculate the sum of the higher reciprocal figurate numbers by means of this method, since these can be obtained as triples, quadruples etc. of the iterated differences of the harmonic series. Already before the end of the year 1672 Leibniz had prepared a paper (A III 1 No. 2) for the *Journal des Sçavans*, but unfortunately at that time publication of the journal was interrupted for more than a year.

3.1   Indirect Reception

In January 1673, Leibniz travelled with a diplomatic delegation of the court of the prince elector of Mainz to London[48]. On 12 February 1673, he visited Robert Boyle and met the mathematician John Pell. During a conversation with Pell, Leibniz mentioned his difference method. Pell declared that such a method already had already been found by François Regnauld and had been published, in 1670, in a book

---

[45] Leibniz (1920, 87); "Pleraque theoremata Geometriae indivisibilium quae apud Cavalerium, Vincentium, Wallisium, Gregorium, Barrovium extant statim ex calculo patent" (A VII 5 No. 44, 313.).

[46] "Secuti hos sunt Jacobus Gregorius Scotus, & Isaacus Barrovius Anglus, qui praeclaris in hoc genere theorematibus scientiam mire locupletarunt." (Leibniz (1686, 104; *GM* V 232.)).

[47] Huygens (1888–1950, vol XIV, 144–150).

[48] See Hofmann (1974, 23–35).

by Gabriel Mouton[49]. Leibniz was able to consult the book the following day when visiting Henry Oldenburg, the secretary of the Royal Society. He wrote immediately a short defence and a presentation of his method for the Royal Society, in which he also stated his results of the summation of the reciprocal figurate numbers (A III 1 No. 4, 29). On 20 February 1673 he submitted a request for becoming a member of the Royal Society, including a paper with his results (A III 1 No. 7$_2$). This letter was read at the meeting of the Royal Society on 1 March 1673.

Since Leibniz was already back in Paris on 8 March 1673, he must have left London about two weeks earlier, one week before the meeting took place during which his paper was read. Shortly before his departure from London, Leibniz must have been informed about a certain reaction to his short defence, because in a letter dated 8 March 1673 he asked Oldenburg for more detailed information on Pell's comments. This was the occasion on which he stated the priority of Mengoli concerning the summation of the reciprocal triangular numbers (A III 1 No. 9, 43; OC IX No. 2165, 491). Leibniz received this fuller account of Pell's reaction in a large letter from Oldenburg dated 20 April 1673, which also informed him of his successful election into the Royal Society (A III 1 No. 13; OC IX No. 2196, 2196a, 2202): Oldenburg writes that John Collins told him that Mengoli's result had been published in his book entitled *De additione fractionum sive quadraturae arithmeticae*, Bologna 1658 (recte: Mengoli 1650). There Mengoli indicates that he found the sum of the reciprocal figurate numbers, but failed—as he himself admitted—to find the sum of the series of the reciprocal square numbers and the sum of the harmonic series (A III 1 No. 13$_2$, 60; OC IX No. 2196, 557). Leibniz replies to Oldenburg on 26 April 1673 (and a second time on 24 May, erroneously believing that his first letter gone missing), writing that he has not yet been able to consult Mengoli's book (A III 1 No. 17, 88, and No. 20, 92–93; OC IX No. 2208, 596 and No. 2233, 648). Since Leibniz also assumed that Mengoli had only found the sums of finite series, Oldenburg makes clear (again with the help of Collins) that Mengoli had actually found the sum of the infinite series of the reciprocal figurate numbers and had demonstrated that the harmonic series cannot be summed, since it exceeds any finite value (A III 1 No. 22, 98; OC IX No. 2238, 667)[50].

The exchange of letters between Leibniz and Oldenburg was interrupted thereafter for 1 year; Leibniz then writes to Oldenburg in July 1674, in order to inform him of the progress he made in the construction of the calculating machine and of his new results concerning quadratures (circle series, cycloid segment: A III 1 No. 30; OC XI No. 2511). Mengoli is not mentioned again until Leibniz's letter to Oldenburg of 16 October 1674, in which he reports on the number theoretical controversy between Jacques Ozanam and Mengoli concerning the six-square problem (A III 1 No. 35 128–129; OC XI No. 2550, 98–99). Mengoli had sought to prove that the problem posed by Ozanam was unsolvable and published this proof, unaware of the serious errors it contained. Ozanam subsequently took delight in humiliating Mengoli by reprinting his flawed proof together with his own successful numerical

---

[49] Mouton (1670, 384).

[50] For accounts of Mengoli's results and methods see Giusti (1991), Massa (1997), Massa Esteve (2006), Massa Esteve/Delshams (2009) .

solution to the problem[51]. Leibniz who—unlike James Gregory—apparently was not able to solve the problem himself does not seem to have disregarded the mathematical abilities of Mengoli because of this error. In his copy of Ozanam's final flyleaf he only points to the places of error (A VII 1 No. 39, 236–237) and in a sheet enclosed he simply records the mere facts: "Mengoli was wrong, and the example shows that it is possible"[52]. In one of his designs for an international science organization, *Consultatio de naturae cognitione* [1679], Leibniz mentions Mengoli among the scholars, whose cooperation he desires (A IV, 3 No. 133, 868).

Evidently, Leibniz did not get access to Mengoli (1650) during his stay in Paris. But, when he visited London for a second time in October 1676, he made excerpts relating to Mengoli from the correspondence between James Gregory and John Collins. In the sections copied by Leibniz there is a passage on Mengoli's proof of the divergence of the harmonic series, characterized by Leibniz in a marginal note as "ingenious" ("ingeniose", A III 1 No. 88_2, 486–487).

3.2   Leibniz's Excerpts from Mengoli's Circolo

It has been known since the 1920s that Leibniz in April 1676 finally had the opportunity to study Mengoli's book *Circolo* (1672) and that he made extensive excerpts from this work[53]. According to the *Catalogue critique* of the manuscripts of Leibniz (Rivaud (1914–1924), quoted as Cc 2), the first part of these three excerpts (Cc 2, No. 1383 A, 1383 B, 1384) is missing, but probably this missing item is at least partly identical with the manuscript LH 35 XII 1 fol. 9–10 (=Cc 2, No. 1398, 1400, 1401), entitled *Arithmetica infinitorum et interpolationum figuris applicata*, and printed in A VII 3 No. 57_2. This had formerly been located together with the excerpts by Leibniz (see A VII 3 No. 57_1) and had been removed to different place within the collection of Leibniz's manuscripts at an unknown date before the end of the 19th Century.

In *Arithmetica infinitorum et interpolationum figuris applicata* Leibniz essentially discusses the triangular tables of Mengoli (1672, 3–10), and tries to find a method for the computation of the partial sums of the harmonic series. With the help of these tables Mengoli determines by interpolation areas of curve segments, something he did already in Mengoli (1659). The values in these tables represent

---

[51] Mengoli suffered hard from this failure as is obvious from his letters to A. Magliabecchi vom 1 June 1674 and to A. Marchetti from 2 June 1674, published in Mengoli (1986, 41–44).

[52] "Erravit Mengolus, idque possibile esse docet... exemplum" (A VII 1 No. 40, 241). – The affair has been studied in detail by Nastasi/Scimone (1994). Leibniz had communicated Ozanam's problem in the aforementioned letter from 8 March 1673 to Oldenburg (III 1 No. 9, 42; OC IX No. 2165, 490–491). His own contributions from 1672 to 1676 and some of the material by Mengoli and Ozanam is published in A VII 1 No. 37–40, 42–44, 49–52, 55–61, 93, 96–100; see also Hofmann (1969).

[53] A VII 6 No. 13, 113–131; the excerpts are recorded in Cc 2 and are mentioned in Mahnke (1926), 8 n. 7.

Fig. 8 Leibniz's harmonic triangle (End-1673—Mid-1674), A VII 3 No. 30, 337 (detail)

$$
\begin{array}{cccccc}
 & & \cdots & & & \\
 & \frac{1}{5} & \cdots & & & \\
 \frac{1}{4} & \frac{1}{20} & \cdots & & & \\
 \frac{1}{3} & \frac{1}{12} & \frac{1}{30} & \cdots & & \\
 \frac{1}{2} & \frac{1}{6} & \frac{1}{12} & \frac{1}{20} & \cdots & \\
 \frac{1}{1} & \frac{1}{2} & \frac{1}{3} & \frac{1}{4} & \frac{1}{5} & \cdots \\
 \frac{3}{2} & \frac{5}{6} & \frac{7}{12} & \frac{9}{20} & \cdots & \\
 \cdots & \cdots & \cdots & \cdots & &
\end{array}
$$

special values of the beta function in today's terminology[54]. Already after reading the first pages of the work of Mengoli, Leibniz became convinced of the truth of the statements made by Oldenburg (or Collins) in regard to the results of Mengoli concerning the summation of the reciprocal figurate numbers. In addition he was able to recognize that Mengoli had already been in possession of the harmonic triangle, used by Leibniz in several different forms and arrangements since the end of 1672 (e.g. A III 1 No. 2, 9; A VII 3 No. 53_2, 710) and for which Leibniz coined the expression "harmonic triangle" in his manuscript *De triangulo harmonico* (A VII 3 No. 30, 337) between the end of 1673 and the middle of 1674[55]. There Leibniz arranges the terms starting from a horizontal line with the terms of the harmonic sequence and indicates the differences of two neighbouring terms in the lines above, the sums in the lines below (Fig. 8).

Mengoli uses brackets in writing the denominator of a fraction behind the nominator (Fig. 9). Leibniz in his excerpt reproduces this arrangement, but (as in his other manuscripts) uses the common notation for fractions (Fig. 10).

The excerpts from Mengoli's *Circolo* consist of a large sheet with three triangular tables (Mengoli (1672, 16, 19 and 7)), which was later folded (A VII 6 No. 13_1, 113–120), and a folded sheet that bears the title *Pars secunda excerptorum ex Circulo Mengoli et ad eum annotatorum* (A VII 6 No. 13_2, 120–131), containing a text primarily concerned with the circle calculation of Mengoli, starting from page 23 of *Circolo*. This part of the manuscript is, however, partly damaged and can only be deciphered with difficulty; nonetheless, it can can be said that Leibniz reconstructed step by step the most important stages of Mengoli's argumentation. Only in two places did the Italian text cause difficulties for him, with the result that he did not attempt to provide a Latin paraphrase of the content, but instead quoted the

---

[54] See M. Rosa Massa Esteve (2006); a detailed account of the content of Mengoli (1672) is provided in Massa Esteve/Delshams (2009) .

[55] See A VII 3 No. 30, 336–340, and Probst (2006).

Fig. 9 Triangular table, Mengoli (1672, 4)

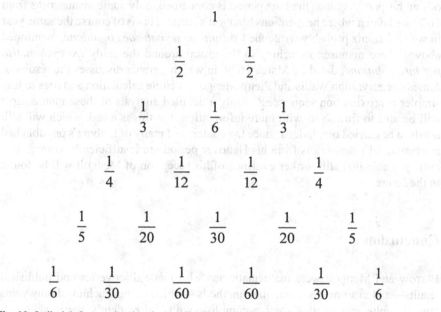

Fig. 10 Leibniz's harmonic triangle (April 1676), A VII 3 No. 57_2, 736

Italian text verbatim[56]. In some places Leibniz inserted comments. For example, he noted in the excerpt that from the approximation sequences for the proportion of the square to the inscribed circle

$$\frac{3}{2} > \frac{3.3.5}{2.4.4.} > \frac{3.3.5.5.7}{2.4.4.6.6} > \dots > \frac{4}{\pi} \dots > \frac{3.3.5.5.7.7}{2.4.4.6.6.8} > \frac{3.3.5.5}{2.4.4.6} > \frac{3.3}{2.4}$$

---

[56] A VII 6 No. 13_1, 129 and No. 13_2, 131; cf. Mengoli (1672), § 105, 39 and § 159, 59.

by forming differences new circle series can be obtained. In the computation of the series

$$\frac{\pi}{4} = \frac{1}{2} + \frac{1}{6} + \frac{2}{45} + \frac{32}{1575} \text{ etc. and } 1 - \frac{\pi}{4} = \frac{1}{9} + \frac{8}{225} + \frac{192}{11025} \text{ etc.}$$

however, minor errors of calculation came about (A VII 6 No. 13_2, 121–125).

It seems that Leibniz did not enter into a further investigation of Mengoli's methods after making the excerpts. In the manuscript of *De quadratura arithmetica*, on which he probably worked until shortly before his departure from Paris at the beginning of October 1676, Leibniz included his results on the summation of the reciprocal figurate numbers and the harmonic triangle without so much as mentioning Mengoli – at least as far as can be established from the extant manuscripts[57]. Up to now no additional documents have been found in Leibniz's manuscripts from his Paris sojourn which provide evidence of further occupation with Mengoli's methods on his part. As far as the later period is concerned, only some manuscripts from 1679 are known where he mentions Mengoli's name. This is of course the same year in which Leibniz probably wrote the *Consultatio de naturae cognitione*, mentioned above. These manuscripts belong to the group around the study *De cyclometria per interpolatione*, dated 26 March 1679, in which Leibniz discusses the results of James Gregory, John Wallis and Pietro Mengoli in circle calculation and tries to find simpler approximation sequences[58]. Only a detailed analysis of these manuscripts will be able to furnish us with more information, but this is a task which will still needs to be carried out. Indeed, since the contents of many of Leibniz's unpublished mathematical manuscripts from his Hanover period are insufficiently known, it is quite possible that still further evidence of his reception of Mengoli will be found in the future.

Conclusion

Barrow and Mengoli were mathematicians who made discoveries and published results—and in some cases also their methods—which Leibniz achieved only years later. Despite appreciating their accomplishments, he evidently never acknowledged any influence of their writings on his own discoveries between 1672 and 1676. He behaved remarkably differently in the cases of Brouncker, Cavalieri, Descartes, Fabri, Fermat, Galileo, Guldin, van Heuraet, Huygens, Pascal, Ricci, Rober-

---

[57] Mengoli is not mentioned in the draft, which Leibniz took to Hanover, the version of the manuscript which he had left in Paris, intended for publication, was lost later; cf. G. W. Leibniz, *De quadratura arithmetica* (A VII 6, introduction, xxi-xxiv, and No. 51, 606–611).

[58] The manuscript of *De cyclometria per interpolatione* is located in LH 35 II 1 fol. 68–73 between the excerpts from Mengoli fol. 67+79, 74,75; fol. 76 discusses the interpolation result of Wallis, fols 77 and 78 contain the sequences for circle approximation and triangular tables from the excerpts from Mengoli's *Circolo*.

val, Saint-Vincent, Sluse, Wallis, and Wren. And this is just to name contemporary authors, who were in his view probably the most important for his mathematical development. One possible reason could be that Leibniz with regard to Mengoli and Barrow was always convinced that he had acquired his knowledge independently of them. A similar picture emerges when we look at the authors Leibniz commonly named as sources and predecessors, for also in these cases he showed different attitudes in different issues. For example, in *De quadratura arithmetica* as well as in the draft of a historical introduction to this treatise, he emphasized the originality of his proof of the method of quadrature of the higher parabolas and hyperbolas contrasted with the results of Fermat and Wallis[59]. He appears to behave the same way in his references to James Gregory and Newton. Informed by Huygens that the auxiliary curve which he used for his circle quadrature had already appeared in print in Gregory (1668), Leibniz added a note in his treatise of the circle quadrature of October 1674, emphasizing his independence of Gregory: "I further do not conceal that Mons. Hugens brought to my attention, to wit that Mr Gregory hit upon the anonymous curve I use here, but for a different purpose, and without perceiving that property which served as the basis for my demonstration"[60]. Later, in *De quadratura arithmetica* of 1676, no such remarks can be found. Also in the sections concerning the circle series there is no mention of Gregory, although Leibniz had already been informed in April 1675 of Gregory's identical series by Oldenburg[61]. Furthermore, the same is true for the sine series of Newton, contained in the same letter from Oldenburg, and later reported again to Leibniz by Georg Mohr[62]. Leibniz did not mention it in his treatment of the sine series, while he praised the binomial theorem of Newton, of which he had gained knowledge through another letter from

---

[59] *De quadratura arithmetica*, A VII 6 No. 51, 588–589. Leibniz cancelled this and other historical sections in the surviving mansucripts of his treatise in order to include them in an ample introduction. There exist several manuscripts, one with outlines of this introduction (A VII 6 No. 39, 427–432), three shorter pieces (A VII 6 No. 40, 433–436; No. 41, 437–439; No. 49_2, 514–518), and an extensive elaboration of the main part, entitled *Dissertatio exoterica de usu geometriae, et statu praesenti, ac novissimis ejus incrementis* (A VII 6 No. 49_1, 483–514; for the remarks concerning Fermat and Wallis see 507). This manuscript is split into two parts preserved in different locations of the Leibniz papers, and both have been published by C. I. Gerhardt separately without recognizing the connection between them (GM V 316–326, and Gerhardt (1891, 157–176, text 167–175). The first part is also printed in A VI 3 No. 54_1, 437–450; a partial translation of the second part is in Leibniz (1920, 186–190), the remarks mentioned are omitted there. The edition of the two isolated fragments has caused misunderstandings. At the end of the text (A VII 6 No. 49_1, 510–514), Leibniz presents briefly the main result of his *Quadratura arithmetica*; Child in Leibniz (1920, 190), declared his incomprehension: "It is difficult to see the object Leibniz had in writing this long historical prelude to an imperfect proof of his arithmetical quadrature, unless it can be ascribed to a motive of self-praise."

[60] "Je ne dissimule non plus que ce Mons. Hugens m'a fait remarquer, sçavoir que Mons. Gregory a touché la Courbe Anonyme dont je me sers icy, mais pour un tout autre usage, et sans s'appercevoir de cette proprieté qui a servi de fondement à ma demonstration" (A III 1 No. 39_2, 169).

[61] A III 1 No. 49_2, 235; OC XI No. 2642, 267.— For Newton's sine series see A III 1, 233; OC XI, 266.

[62] A III 1 No. 80_1, 375; OC XII No. 2893, 268–269; VII 6 No. 17, 162; VII 6 No. 47, 465.

Oldenburg in August 1676[63]. It is not certain if Leibniz discovered the sine series independently, although all the preconditions for him to deduce it in a way analogous to the method he employed for the logarithmic series in proposition XLVII, were given[64]. In *De quadratura arithmetica* he announced the corresponding proposition XLVIII quite ambiguously, relating only to the proof, not to the invention of the series: "Hence a similar rule for the trigonometric regress, or the invention of the sides from the given angles, was not difficult to demonstrate."[65] To sum up, then, from the examples investigated in this essay a common pattern can be established: When Leibniz was convinced that he had discovered a result or a method by himself, he regarded it as his own achievement for which he had no need to acknowledge a debt to any predecessor.

Acknowledgements I am indebted to the editors for helpful criticisms and comments on a previous version of this paper (especially to Philip Beeley for correcting my English), and to an anonymous referee for helpful suggestions.

References

Archimedes. 1675. *Archimedis opera: Apollonii Pergaei conicorum libri IIII. Theodosii sphaerica: methodo nova illustrata, et succincte demonstrata* (Ed. & Transl: I. Barrow). London: Godbid.

Aristotle. *De anima*.

Barrow, Isaac. 1669. *Lectiones XVIII* [...] *in quibus opticorum phaenomenωn genuinae rationes investigantur, ac exponuntur*. London: Godbid.

Barrow, Isaac. 1670. *Lectiones geometricae; in quibus (praesertim) generalia curvarum linearum symptomata declarantur*. London: Godbid.

Barrow, Isaac. 1672. *Lectiones XVIII...in quibus opticorum phaenomenωn genuinae rationes investigantur, ac explicuntur. Annexae sunt Lectiones aliquot geometricae*. London: Godbid.

Barrow, Isaac. 1674. *Lectiones opticae et geometricae*. London: Godbid.

Barrow, Isaac. 1916. *The geometrical lectures of Isaac Barrow* (Ed. & trans: J. M. Child). Chicago: Open Court.

Blank, B. E. 2009 Review of J. S. Bardi: The Calculus wars. *Notices of the AMS* **56**:602–610.

Breger, Herbert. 2004. Ebenen der Abstraktion: Bernoulli, Leibniz und Barrows Theorem. In *Form, Zahl, Ordnung*, ed. R. Seising, 193–202. Stuttgart: Steiner.

Brown, R. C. 2012. *The tangled origins of the Leibnizian calculus*. Hackensack: World Scientific.

Euclid. 1655. *Euclidis elementorum libri XV. breviter demonstrata* (Ed. & trans; I. Barrow). Cambridge: Nealand.

Feingold, M. 1993. Newton, Leibniz and Barrow too, an attempt at a reinterpretation. *Isis* **84**:310–338.

Gerhardt, C. I. 1848. *Die Entdeckung der Differentialrechnung durch Leibniz*. Halle: Schmidt.

Gerhardt, C. I., eds. 1849–63. *Leibnizens Mathematische Schriften*. Berlin and Halle: Asher and Schmidt; reprint Hildesheim: Georg Olms, 1971. 7 vols; cited by volume and page.

Gerhardt, C. I. 1855. *Die Entdeckung der höheren Analysis*. Halle: Schmidt.

---

[63] A III 1 No. 88$_5$; Newton (1960), 20–32.

[64] See A III 1 No. 89$_1$, 566, and Hofmann (1957).

[65] "Hinc jam similem regulam pro regressu trigonometrico, seu inventione laterum ex angulis datis, demonstrare difficile non fuit" (*De quadratura arithmetica*, A VII 6 No. 51, 642).

Gerhardt, C. I. 1891. Leibniz in London. *Sitzungsberichte der Preußischen Akademie der Wissenschaften* **X**:157–176.

Giusti, Enrico. 1991. Le prime ricerche di Pietro Mengoli: la somma delle serie. In *Proceedings of the international meeting "Geometry and complex variables"*, ed. S. Coen, 195–213. New York: Dekker.

Gregory, James. 1668. *Exercitationes geometricae*. London: Godbid and Pitt.

Guicciardini, Niccolò. 2010. Review of Gottfried Wilhelm Leibniz, Sämtliche Schriften und Briefe. *NTM* **18**:545–549.

Heuraet, Hendrik van. 1659. Epistola de transmutatione curvarum linearum in rectas. In R. Descartes, *Geometria*, ed. Fr. van Schooten, 517–520 (2nd ed., part I). Amsterdam: Elzevir.

Hofmann, Joseph E. 1957. Über Leibnizens früheste Methode zur Reihenentwicklung, *Mitteilungen der Deutschen Akademie der Naturforscher Leopoldina*, 3/3, 67–72.

Hofmann, Joseph E. 1969. Leibniz und Ozanams Problem. *Studia Leibnitiana* **1**:103–126.

Hofmann, Joseph E. 1974. *Leibniz in Paris 1672–1676*. Cambridge: Cambridge University Press.

Huygens, Christiaan. 1673. *Horologium oscillatorium*. Paris: Muguet.

Huygens, Christiaan. 1888–1950. *Oeuvres complétes*. La Haye: Nijhoff. 22 vols.; cited by volume and page.

Leibniz, Gottfried Wilhelm. 1686. De geometria recondita et analysi indivisibilium atque infinitorum. *Acta Eruditorum* 5:292–300 (GM V 226–233).

Leibniz, Gottfried Wilhelm. 1693. Supplementum geometriae dimensoriae. *Acta Eruditorum* 12:385–392 (GM V 294–301; a relevant excerpt is available in English translation in D. J. Struik, *A Source Book in Mathematics 1200–1800*, Cambridge (MA), 282–284).

Leibniz, Gottfried Wilhelm. 1920. *The early mathematical manuscripts of Leibniz*. Trans: J. M. Child. Chicago Open Court.

Leibniz, G. W. 1923. *Sämtliche Schriften und Briefe*, ed. Prussian Academy of Sciences (and successors); now: Berlin-Brandenburg Academy of Sciences and Academy of Sciences zu Göttingen. 8 series, Darmstadt (subsequently: Leipzig); now: Berlin: Otto Reichl (and successors); now: Akademie Verlag; cited by series, volume, number and page.

Leibniz, Gottfried Wilhelm. 1993. *De quadratura arithmetica circuli ellipseos et hyperbolae cujus corollarium est trigonometria sine tabulis*. Critical edition with commentary by E. Knobloch, Göttingen: Vandenhoeck & Ruprecht. (Reprinted in 2004 with French translation by M. Parmentier: *Quadrature arithmétique du cercle, de l'ellipse et de l'hyperbole*, Paris: Vrin.).

Mahnke, Dietrich. 1926. *Neue Einblicke in die Entdeckungsgeschichte der höheren Analysis*. Berlin: Akademie der Wissenschaften. (Abhandlungen der Preuß. Phys.-math. Klasse, Jahrgang 1925, Nr. 1.).

Mahoney, Michael S. 1990. Barrow's mathematics: Between ancients and moderns. In *Before Newton*, ed. M. Feingold, 179–249. Cambridge: Cambridge University Press.

Massa Esteve, M. Rosa. 1997. Mengoli on 'Quasi proportions'. *Historia Mathematica* **24**:257–280.

Massa Esteve, M. Rosa. 2006. Algebra and geometry in Pietro Mengoli (1625–1686). *Historia Mathematica* **33**:82–112.

Massa Esteve, M. Rosa, and Amadeu Delshams. 2009. Euler's beta integral in Pietro Mengoli's works. *Archive for History of Exact Sciences* **63**:325–356.

Mayer, Uwe. 2006. Mündliche Kommunikation und schriftliche Überlieferung: die Gesprächsaufzeichnungen von Leibniz und Tschirnhaus zur Infinitesimalrechnung aus der Pariser Zeit. In *VIII International Leibniz Congress, Unity in Plurality*, eds. H. Breger, J. Herbst, and S. Erdner, pp, 588–594. Hanover: Gottfried Wilhelm Leibniz Gesellschaft.

Mengoli, Pietro. 1650. *Novae quadraturae arithmeticae*. Bologna: Monti.

Mengoli, Pietro. 1659. *Geometriae speciosae elementa*. Bologna: Ferreni.

Mengoli, Pietro. 1672. *Circolo*. Bologna: Herede del Benacci.

Mengoli, Pietro. 1986. *La corrispondenza di Pietro Mengoli*. In eds. G. Baroncini and M. Cavazza. Florence: Olschki.

Metius, Adriaan. 1625. *Geometria practica*. Franeker: Balck.

Metius, Adriaan. 1633. *Manuale arithmeticae et geometriae practicae*. Amsterdam: Laurentsz.

Mouton, Gabriel. 1670. *Observationes diametrorum solis et lunae apparentium*. Lyon: Liberal.

Nastasi, P., and A. Scimone. 1994. Pietro Mengoli and the six-square problem. *Historia Mathematica* **21**:10–27.

Newton, Isaac. 1960. The Correspondence, ed. H. W. Turnbull, J. F. Scott, A. R. Hall, vol. 2. Cambridge: CUP.

Nauenberg, Michael. 2014. Barrow, Leibniz and the geometrical proof of the fundamental theorem of the calculus. *Annals of Science* 71:335–354.

Oldenburg, Henry. 1965–86. *The correspondence*, ed. A. R. Hall and M. Boas Hall. Madison and London: Univ. of Wisconsin Press, Mansell, Taylor & Francis. 13 vols.; cited by volume and number or page.

Probst, Siegmund. 2006. Differenzen, Folgen und Reihen bei Leibniz (1672–1676). In *Wanderschaft in der Mathematik,* eds. M. Hyksová and U. Reich, 164–173. Augsburg: Rauner.

Probst, Siegmund. 2011. Leibniz' Lektüre von Barrows 'Lectiones geometricae' in den Jahren 1673–16764. In *IX International Leibniz Congress, 'Nature and Subject',* eds. H. Breger, J. Herbst, and S. Erdner, pp, 869–877. Hanover: Gottfried Wilhelm Leibniz Gesellschaft.

Rivaud, Albert, et al. 1914–1924. *Catalogue critique des manuscrits de Leibniz. Fascicule II (Mars 1672—Novembre 1676)*. Poitiers: SFIL. (quoted as: Cc 2).

Sluse, René-François de. 1673. An extract of a letter […] to the Publisher […] concerning his short and easie method of drawing tangents to all geometrical curves. *Philosophical Transactions* **VII**:5143–5147.

Turnbull, H. W., ed. 1939. *James Gregory tercentenary memorial volume*. London: Bell.

Wahl, Charlotte. 2011. Assessing mathematical progress: Contemporary views on the merits of Leibniz's infinitesimal calculus. In *IX International Leibniz Congress, 'Nature and Subject',* eds. H. Breger, J. Herbst, and S. Erdner, 1172–1182. Hanover: Gottfried Wilhelm Leibniz Gesellschaft.

Wallis, John. 1672. Epitome binae methodi tangentium. *Philosophical Transactions* **VII**:4010–4016.

Witt, Jan de. 1659. Elementa curvarum linearum. In R. Descartes, *Geometria,* ed. Fr. van Schooten, 153–340 (2nd ed., part II). Amsterdam: Elzevir.

Part III
The Problem of Infinity

Leibniz's Actual Infinite in Relation to His Analysis of Matter

Richard T. W. Arthur

1 Introduction: The Actually Infinite Division of Matter

It is well known that Leibniz held that matter is infinitely divided, and that there are infinitely many monads. But the connection between these two theses has not been well understood, and this has led to perplexity about Leibniz's views on the actual infinite and on the composition of matter, and also prompted accusations of inconsistency. Georg Cantor, for example, seized on Leibniz's endorsement of the actual infinite as an important precedent for this theory of the transfinite. Acknowledging that Leibniz "often pronounces himself against infinite number", Cantor declared that he was nevertheless "in the happy position of being able to cite pronouncements by the same thinker in which, to some extent in contradiction with himself, he expresses himself unequivocally *for* the actual infinite (as distinct from the Absolute)."[1] He quotes a typical passage to this effect from Leibniz's letter to Simon Foucher in 1693:

> I am so much in favour of the actual infinite that instead of admitting that Nature abhors it, as is commonly said, I hold that she manifests it everywhere, the better to indicate the perfections of her Author. Thus I believe that there is no part of matter which is not, I do not say divisible, but actually divided; and consequently the least particle ought to be considered as a world full of an infinity of different creatures.[2]

It should be remembered that Cantor was still battling with an Aristotelian orthodoxy in the philosophy of mathematics, a fact that explains the importance for him of Leibniz's claim that there are *actually infinitely many* created substances

---

[1] Cantor, "Grundlagen einer allgemeinen Mannigfaltigkeitslehre" (1883), in Cantor 1932, p. 179. All translations in the present paper are my own. I cite current English translations where they exist for ease of reference.

[2] Letter to Foucher, *Journal de Sçavans*, August 3, 1693, GP I 416 [=A II 2, N. 226. Leibniz an Simon Foucher (Wolfenbüttel, Ende Juni 1693), p. 713.].

R. T. W. Arthur (✉)
Department of Philosophy, McMaster University, Hamilton, ON, Canada
e-mail: rarthur@mcmaster.ca

© Springer Netherlands 2015
N. B. Goethe et al. (eds.), *G.W. Leibniz, Interrelations between Mathematics and Philosophy,* Archimedes 41, DOI 10.1007/978-94-017-9664-4_7

(monads), in defiance of the Aristotelian stricture that the infinite can exist only *potentially*.

Cantor, of course, believed that if there are actually infinitely many discrete creatures in any given part of matter, then there must be a corresponding infinite *number* of them. Indeed, his theory of transfinite numbers was designed to enable one to express precisely such statements. Now one might think that his success in establishing that mathematical theory would have marked the end of his interest in Leibniz's philosophy of matter and monads. That, however, is far from the case. In a side of his thought that perhaps deserves to be better known, Cantor did in fact use his theory of transfinite numbers to express such statements about the actually infinitely many constituents of matter. Declaring himself a follower of Leibniz's "organic philosophy", Cantor held "that in order to obtain a satisfactory *explanation of nature*, one must posit the ultimate or properly *simple* elements of matter to be *actually infinite in number*."[3] He continued:

> In agreement with Leibniz, I call these simple elements of nature *monads* or *unities*. [But since] there are *two specific, different types of matter interacting with one another*, namely *corporeal matter* and *aetherial matter*, one must also posit two different classes of *monads* as foundations, *corporeal monads* and *aetherial monads*. From this standpoint the question is raised (a question that occurred neither to Leibniz nor to later thinkers): what *power* is appropriate to these types of matter with respect to their elements, insofar as they are considered *sets* of *corporeal* and *aetherial monads*. In this connection, I frame the hypothesis that the *power [Mächtigkeit]* of the corporeal monads is (what I call in my researches) the *first* power, whilst the power of aetherial matter is the *second*. (Cantor 1932, pp. 275–276)

That is, Cantor posits corporeal monads which, as discrete unities, are equinumerous with the natural numbers, and therefore have a power or cardinality $\aleph_0$, the first of the transfinite cardinal numbers. He further proposes that the aether, which he assumes to be continuous, is composed of aetherial monads equinumerous with the points on a line, i.e. that the number of aetherial monads is equal to $\aleph_1$, the second cardinal number, which he believed (but was never able to prove) to be the power of the continuum.[4]

There were certainly precedents for representing Leibniz's monads in this way as *elements* of matter. Euler, for instance, interpreted Leibniz's monads as "ultimate particles which enter into the composition of bodies" (Euler 1843, p. 39).[5]

---

[3] This and the quotations following are culled from Cantor's "*Über verschiedene Theoreme aus der Theorie der Punktmengen in einem n-fach ausgedehnten stetigen Raume G_n*" (1885), (Cantor 1932, pp. 275–276).

[4] Cantor 1932, p. 276. Cf. Dauben 1979, p. 292. Since Cantor's time it has been shown that his Continuum Hypothesis is consistent with (by Gödel, in 1940) but independent of (by Cohen, in 1963) Zermelo-Fraenkel set theory, the standard foundation of modern mathematics, provided ZF set theory is consistent.

[5] Euler also called Leibniz's monads *parts* of bodies that result from a "limited division" (1843, pp. 48–49). Perhaps also contributing to these eighteenth century misunderstandings of Leibniz were the views of Maupertuis, where the fundamental particles of matter are physical points endowed with appetition and perception, and of Boscovic, whose particles are interacting point-sources of attractive and repulsive force.

But modern commentators, equipped with a much more comprehensive selection of Leibniz's writings, would not consider Cantor's reinterpretation of Leibnizian monads a faithful elaboration of his views. For Leibniz conceived his monads or simple substances neither as *interacting* with one another, nor as *elements* out of which matter is composed. Both these ideas are closer to the views of Christian Wolff, who advocated a theory of simple substances that were avowedly not monads in Leibniz's sense. Instead, Wolff's substances were supposed to give rise to material atoms through their interactions, which atoms were the elements out of which extended bodies were then aggregated.[6]

Putting those objections to one side, however, modern scholars have by and large taken Cantor's side on the actual infinite, agreeing with his criticisms of Leibniz's rejection of infinite number. They point out that Leibniz would have had a precedent for embracing infinite number in Galileo Galilei, who in his *Two New Sciences* had declared matter to be composed of an actually infinite number of atoms, separated by infinitely small voids. Had Leibniz followed Galileo's lead, it is often asserted, he might have well anticipated Cantor's transfinite.[7]

I am not convinced, however, that such claims are justified. Leibniz insisted that the axiom that the whole is greater than its (proper) part must hold in the infinite as well as in the finite, whereas Cantor followed Dedekind in maintaining that an infinity of elements could be equal in number to an (infinite) proper subset of those elements, and consequently denied the applicability of the part-whole axiom to infinite sets. Galileo, by contrast, denied that the notions of being 'greater', 'equal to' or 'less than' applied in the infinite at all, thus contradicting both Leibniz and Cantor. But this very denial made his approach unsuitable as a foundation for the mathematics of the infinite.[8]

---

[6] Wolff's simple substances should not, however, be conflated with the (extended) primitive corpuscles which he supposed were constituted by their interactions. For whereas these primitive corpuscles are finitely extended and only physically indivisible, Wolff's simple substances are unextended because partless, and aggregate into extended bodies. See (Wolff 1737, § 182, 186, 187). A succinct and accurate summary of these aspects of Wolff's philosophy is given by Matt Hettche in his (2006).

[7] Rescher states that the Cantorian theory of the transfinite, point-set topology and measure theory "have shown that Leibniz's method of attack was poor. Indeed, Galileo had already handled the problem more satisfactorily…" (Rescher 1967, p. 111); Gregory Brown asserts that, "had he not jumped the gun in rejecting the possibility of infinite number and infinite wholes, Leibniz, having already surmounted the prejudice against actual infinities, would have been well placed to anticipate the discoveries of Cantor and Frege by at least 200 years." (Brown 2000, p. 24; see also his 1998, pp. 122–123).

[8] For Galileo the infinitely small parts of the continuum are *non quante*, whereas Leibniz insisted from the beginning that they were quantifiable. Leibniz continued to maintain that the infinite and the infinitely small are quantifiable even after defining them as useful fictions, which is why they are treatable in his differential calculus. I have argued this at greater length in Arthur 2001.

2   The Fictionality of Infinite Wholes and Collections

Leibniz, in fact, was well aware of Galileo's argument in the *Two New Sciences* that "in the infinite there is neither greater nor smaller", having himself made notes on it in the Fall of 1672 (A VI 3, 168; LoC 6–9). He gives Galileo's demonstration as follows:

> Among numbers there are infinite roots, infinite squares, infinite cubes. Moreover, there are as many roots as numbers. And there are as many squares as roots. Therefore there are as many squares as numbers, that is to say, there are as many square numbers as there are numbers in the universe. Which is impossible. Hence it follows either that in the infinite the whole is not greater than the part, which is the opinion of Galileo and Gregory of St. Vincent, and which I cannot accept; or that infinity itself is nothing, i.e. that it is not one and not a whole. (LoC 9)

Leibniz had made these notes soon after his arrival in Paris, by which time he was already at work on infinite series under Huygens' guidance. These studies in proto-calculus did nothing to dissuade him from this opinion about the infinite not being a true whole. In fact, the expressions he found for the area under a hyperbola in terms of infinite series gave a pictorial representation of an infinite whole being equal to its part, and at the same time seemed to him to confirm the fictional nature of such a whole.

The following is a condensed version of Leibniz's argument from a paper written in October 1674 (A VII 3, 468). He gives a symmetrical diagram (Fig. 1 below) of a hyperbola with centre A, vertex B, and radius $AC=BC=a$, which, without loss of generality, may be set equal to 1. M represents the fictional point where the curve "meets" each line AF... at infinity. Here the x-axis is the line ACF... across the top of the figure, the y-axis runs with increasing y down the page from A through DCH to F, and the dotted lines DE and HL represent the variable abscissa x. Thus $AD=AC-CD=1-y$, and $AH=AC+CH=AC+CD=1+y$.

Fig. 1 Leibniz's hyperbola

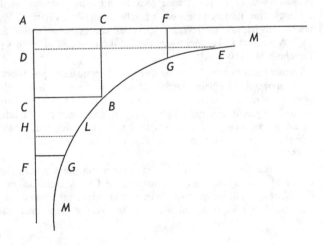

Leibniz now sets about discovering the area under the curve between the horizontal line CB and the x-axis running across the top of the figure. We have $DE = 1/AD = 1/(1-y)$, which he expands as an infinite series:

$$DE = 1/(1-y) = 1 + y + y^2 + y^3 + y^4 + y^5 + \ldots$$

Now the area in question is obtained by "applying" the variable line DE to the line $AC = 1$, giving

$$\text{Area}(ACBEM) = 1 + \frac{1}{2} + \frac{1}{3} + \frac{1}{4} + \frac{1}{5} + \frac{1}{6} \ldots$$

In modern terms, Leibniz has *integrated* a power series expansion of $1/(1-y)$ with respect to y, $\int(1 + y + y^2 + y^3 \ldots)\, dy$, between 0 and 1. By a similar argument, he obtains

$$\text{Area}(CFGLB) = 1 - \frac{1}{2} + \frac{1}{3} - \frac{1}{4} + \frac{1}{5} - \frac{1}{6} \text{ etc.} = \frac{1}{2} + \frac{1}{12} + \frac{1}{30} + \frac{1}{56} + \frac{1}{90} \ldots$$

Now Leibniz subtracts this finite area $CFGLB$ from the infinite area $ACBEM$, to get

$$\text{Area}(ACBEM - CFGlB) = 1 - (1) + \frac{1}{2} - \left(-\frac{1}{2}\right) + \frac{1}{3} - \left(+\frac{1}{3}\right) + \frac{1}{4} - \left(-\frac{1}{4}\right) + \frac{1}{5} - \left(+\frac{1}{5}\right) \ldots$$

$$= \frac{2}{2} + \frac{2}{4} + \frac{2}{6} + \frac{2}{8} + \frac{2}{10} + \frac{2}{12} \ldots$$

$$= 1 + \frac{1}{2} + \frac{1}{3} + \frac{1}{4} + \frac{1}{5} + \frac{1}{6} \ldots$$

$$= \text{Area}(ACBEM)$$

That is, subtracting the area $CFGLB$—an area that is perfectly definite and observable—from the area under the curve leaves that area the same! Leibniz comments:

> This is remarkable enough, and shows that the sum of the series 1, $^1/_2$, $^1/_3$, etc. is infinite, and therefore that the area of the space $ACGBM$, even when the finite space $CBGF$ is taken away from it, remains the same, i.e. this takes away nothing observable (*notabile*). By this argument it can be concluded that the infinite is not a whole, but a fiction, since otherwise the part would be equal to the whole. (A VII 3, 468)

It is noteworthy here that Leibniz concludes that the sum of the series $1 + ^1/_2 + ^1/_3 + \ldots$ is infinite on the implicit grounds that, if it is a whole, it is equal to a proper part of itself. As is well known, Dedekind, having followed Bolzano in characterizing a set as infinite if it can be set in 1-1 correspondence with a proper subset of itself, defines two sets as equal iff there exists a 1-1 correspondence (bijection) between their elements—the definition of equality of sets that Cantor takes as foundational for his theory of size of infinite sets.[9] As Russell remarks in his *Introduction to Mathematical Philosophy* (1919, p. 81), this enables Cantor to avoid Galileo's Paradox. For if one takes the criterion of bijection as defining the equality ("similarity" in his terminology) of infinite sets, then "there is no contradiction, since an infinite collection can perfectly well have parts similar to itself'. That is, assuming infinite collections and the Dedekindian definition of equality adopted by Cantor, it can no longer be maintained that it is "self-contradictory that 'the part should be equal to the whole'" (80–81). The contradiction is avoided by jettisoning the part-whole axiom (P) rather than the assumption that an infinity of terms can be collected into one whole (C).

As I have argued elsewhere, this is typical of a reductio argument, where the conclusion to draw from proving a set of premises inconsistent depends on which premise one is prepared to reject. Leibniz, insisting on P, infers ¬C. Russell and Cantor, insisting on C, infer ¬P (Arthur 2001, p. 103). Gregory Brown has taken issue with this logic, arguing that, "given the consistency of Cantorian set theory, it would again appear that Leibniz's argument against infinite number and infinite wholes must be unsound." (Brown 2005, p. 481). He argues that if Leibniz's argument were sound, then Cantor's theory could not even be consistent, because it is erected on definitions of 'less than', 'greater than', and 'equal to'—"given in terms of one–one correspondence of sets" (484)—that are in contradiction with Leibniz's. But this is just to claim that, *given C*, P cannot be taken to be axiomatically true.

In mounting his case against Leibniz, Brown appeals to an argument given by Jose Benardete in his book on infinity (1964), who accused Leibniz of equivocating between different senses of the concept of equality—a criticism originating with Russell in his (1919). Benardete's criticism has recently been developed further by Mark van Atten, who argues that "Russell and others have observed that Leibniz's argument is not correct because it rests on an equivocation on the concept of equality." (van Atten 2011, p. 123). Van Atten makes his case by quoting several versions of Leibniz's argument, and then giving a reconstruction in which the third line is "3. The multitude of the squares is equal to a part of the whole of the numbers", and the fifth is "5. The multitude of the squares is equal to the whole of the numbers." "Clearly," he writes, "in line 3 'is equal to' means 'is identical to', while in line 5 it means 'can be put in a bijection with'" (123).

I do not agree that Leibniz is guilty of such an equivocation. In all versions he gives of this argument he accepts that if there is a bijection between the terms in two multiplicities (*multitudines*),[10] then they are equal. That is, Leibniz accepts that

---

[9] Although Bernard Bolzano recognized that it is a characteristic property of an infinite set that its members may be set in 1–1 correspondence with those of a proper subset of itself, he denied that the two sets would therefore have the same number of members. That was Dedekind's contribution. See Mancosu 2009, esp. pp. 624–627 for an illuminating discussion.

[10] Leibniz's neutral term for a plurality is *multitudo*, which Van Atten and others translate literally as "multitude", whereas I follow Russell in translating it as *multiplicity*. Unfortunately, Leibniz is not always consistent in his use of terms, sometimes using *collectio* (collection) as a synonym. But

bijection implies equality; he does not have to accept it as a definition of equality. Van Attenuses against him his definition of sameness or coincidence of terms as those "which can be substituted for each other without affecting the truth" to show that Leibniz uses "equality" in two diverse senses: "the concepts of equality in lines 3 and 5 are diverse or different, as substitution of the one for the other does not preserve truth here" (124). But the only criterion of equality Leibniz appeals to is that of bijection: if the even numbers have a 1-1 correspondence with a subset of the whole numbers, the multiplicity of even numbers is equal to a part of the multiplicity of whole numbers. Here by "part", Leibniz understands "proper part" or proper subset, as van Atten acknowledges, so that one multiplicity B is a part of another multiplicity A if there are no terms in B that are not in A, while there are terms in A that are not in B. It will then follow that, if the multiplicity of natural numbers N and that of square numbers Q can be treated as wholes, Q is a part of the whole N; but since there is a 1-1 correspondence between their terms, they are equal: the whole is equal to the part. In a letter to Justus Christoph Böhmer in June, 1694, Leibniz phrases the argument in exactly this way:

> If any A has a B corresponding to it, and any B has an A corresponding to it; it follows that there are as many A as there are B and vice versa.
> But any number has a square corresponding to it, and any square has a number corresponding to it.
> Therefore there are as many numbers as there are squares, and vice versa.
> Therefore the multiplicity of all numbers (if such there is), is equal to the multiplicity of all squares.
> But the multiplicity of all numbers is a whole, and the multiplicity of all squares is a part, because the multiplicity of all numbers contains numbers that are not squares as well.
> Therefore the whole is equal to the part. Which is absurd.
> Therefore there is no multiplicity or number of all numbers, nor of all squares, but rather such a thing is chimerical. (A II 2B, 814)

Thus, given the part-whole axiom P (and his interpretation of 'part'), Leibniz is correct to infer C, that an infinite collection of terms does not constitute a true whole. And this denial of the existence of infinite collections is equivalent, as accepted on all sides, to the denial of infinite number.

Now it is true that Leibniz claimed that he could *demonstrate* P from other notions, such as "B is B" and "each thing is equal to itself", as Brown reminds us in his criticisms of Leibniz's argument. But the very consistency of Cantor's theory of the transfinite, he argues, shows that Leibniz must have been wrong to claim this (Brown 2005, p. 484).[11] If Leibniz had indeed demonstrated it, then Cantor's theory would be unsound. Exactly this kind of criticism has been subjected to a detailed

---

he wants to deny infinite collections, while still maintaining that there are infinite multiplicities or aggregates: one can collect together all the terms of a multiplicity if it is finite, but not if it is infinite.

[11] Of course, one could also object to Brown that Cantor's own "naive" set theory is known to be *in*consistent; and that while it has now been replaced as the foundation of mathematics by Zermelo-Fraenkel set theory with the axiom of choice (ZFC), the consistency of ZFC (as Brown himself acknowledges on p. 486) has *not* been established yet. It is simply assumed that if ZFC were inconsistent, a contradiction in it would have been discovered by now.

rebuttal by Herbert Breger in his (2008), in response to Van Atten's criticisms. An argument of this kind, Breger argues, can show that Leibniz was wrong *only if one assumes that there is just one correct foundation for mathematics, and that that was given by Cantor*. Such a conception of mathematics, Breger argues, is belied by its actual historical development.[12] It is clear, he argues, that each decision to generalize concepts of mathematics—such as the decision to admit negative numbers, imaginary numbers, or the transfinite—involved a contingent decision, made on philosophical grounds. One can argue that the revised definition is superior on the basis of the mathematics it enables, but it is a historiographical mistake then to interpret *in terms of this revised definition* statements made about numbers prior to the acceptance of this definition, in order to show that these statements are false. On Leibniz's conception, the part-whole axiom is constitutive of number just because no comparisons of quantity can be made without presupposing it. If bijection implies equality, this entails that infinite number entails a contradiction, given P. One cannot find fault with this argument by giving transfinite numbers as a counterexample, since these are numbers only in the Cantorian revised sense of number based on the existence of infinite collections, for which P fails: they are not instances of the numbers Leibniz is discussing.[13]

Breger's point about the contingency of the historical development of mathematics, as well as his and my defence of the tenability of Leibniz's retaining the part-whole axiom, have recently received strong indirect support from an intriguing article by Paolo Mancosu on measuring the size of infinite collections (Mancosu 2009). Mancosu traces the idea that the size of an infinite set can be based on the part-whole axiom from its historical roots to recent developments in mathematics. He uses the existence of alternative theories of class sizes based on the part-whole axiom (due to Katz, Benci, Di Nasso and Forti) to "debunk" arguments such as Kurt Gödel's that the Cantorian conception of infinite number (based on the Dedekindian definition of the equality of sets) is inevitable.[14] These theories of class size, of course, depend on the existence of infinite collections, which Leibniz denies, so they are not compatible with Leibniz's own approach. But their very existence shows that there is nothing inevitable about having to adopt the Dedekindian definition of the equality of sets or about having to reject the part-whole axiom.

---

[12] "One must renounce the assumption … that there is *one* mathematics—an assumption that should in fact have gone out of date with the acceptance of non-Euclidean geometries." (Breger 2008, p. 314).

[13] Cf. Breger's discussion of the two notions of 'being the same number', (i) A = B iff neither A nor B is a proper part of the other, and (ii) A = B iff there is a bivalent mapping between them: "The fact that one finds objects outside the theory examined here for which both notions are not equivalent is of no importance within the theory." (2008, p. 314).

[14] "I have hoped to show that the possibility of comparing Cantor's theory against the alternative theories of class sizes (CS) and numerosities allows us to analyze more finely, and in some cases debunk, the arguments that claim either the inevitability of the Cantorian choice (Gödel) or that account for the (alleged) explanatory nature of the Cantorian generalization by appealing to the (alleged) nonrational nature of preserving the part–whole principle." (Mancosu 2009, p. 642).

Furthermore, despite his rejection of infinite number and infinite collections, Leibniz maintained that one can make true statements about an infinity of terms, such as in asserting that there are infinitely many terms in an infinite series. This does not mean that there is an infinite *number* of terms. It means that there are more terms than can be assigned any finite number. As Leibniz explained to Johann Bernoulliin a letter of February 21, 1699,

> We can conceive an infinite series consisting merely of finite terms, or terms ordered in a decreasing geometric progression. I concede the infinite multiplicity of terms, but this multiplicity forms neither a number nor one whole. It means only that there are more terms than can be designated by a number; just as there is for instance a multiplicity or complex of all numbers; but this multiplicity is neither a number nor one whole. (GM III 575)

This is Leibniz's actual but *syncategorematic* infinite. This term alludes to the distinction first formulated by Peter of Spain, and later elaborated by Jean Buridan, Gregory of Rimini and William of Ockham, who claimed that to assert that the continuum has infinitely many parts in a *syncategorematic* sense is to assert that "there are not so many parts finite in number that there are not more (*partes non tot finitas numero quin plures,* or *non sunt tot quin sint plura*)". The statement is called *syncategorematic* because the term 'infinite' occurs in it, but that term does not actually have a referent corresponding to it.[15] Rather, it gains its meaning from the way the statement as a whole functions. This is contrasted with the *categorematic* sense of infinity, according to which to say that there are infinitely many parts is to say that there is a number of parts greater than any finite number, i.e. that there is an infinite number of parts. As Leibniz elaborates in his *New Essays*,

> It is perfectly correct to say that there is an infinity of things, i.e. that there are always more of them than can be specified. But it is easy to demonstrate that there is no infinite number, nor any infinite line or other infinite quantity, if these are taken to be genuine wholes. The Scholastics were taking that view, or should have been doing so, when they allowed a 'syncategorematic' infinite, as they called it, but not a 'categorematic' one. (Leibniz, *New Essays*, 1981, § 157; GP V 144; Russell 1900, p. 244)

Leibniz's actual but syncategorematic infinite is thus distinct from Aristotle's potential infinite, in that it embraces an infinity of actually existents, but it also differs profoundly from Cantor's theory of the actual infinite as transfinite in that it denies the existence of infinite collections or sets that are the basis of transfinite set theory. It is not the existence of an infinite plurality of terms that is denied, but the existence of an infinite collection, and thus an infinite number, of them. For instance, to assert that *there are infinitely many primes*, is to assert that, for any finite number x that you choose to number the primes, there is a number of primes y greater than this: $\forall x \exists y (Fx \rightarrow y > x)$, where $Fx := \,$'x is finite', and x and y are any numbers. By contrast, to assert their infinity *categorematically* would be to as-

---

[15] See O. B. Bassler's erudite footnote on the syncategorematic and categorematic in his 1998, 855, n. 15, and the references cited therein. See also Sam Levey's 2008 for a careful elaboration of Leibniz's fictionalism.

sert that there exists some one number of primes y which is greater than any finite number x, i.e. that $\exists y \forall x (Fx \rightarrow y > x)$ —i. e. that there exists an infinite number. Interestingly, Euclid's proof that there are infinitely many primes begins by assuming that there finitely many, so that there is a greatest prime, and then deriving a contradiction. But the negation of the assumption that there is a greatest finite prime, $\exists x \forall y [Fx \& (y \neq x \rightarrow y \leq x)]$, is provably equivalent to $\forall x \exists y [Fx \rightarrow (y \neq x \& y > x)]$. This says "for any finite number of primes, there is a number of primes different from and greater than this". This is the syncategorematic actual infinite, not the categorematic, which would be $\exists y \forall x [Fx \rightarrow (y \neq x \& y > x)]$: "there exists a number of primes greater than any finite number".

As I have argued elsewhere (Arthur 2008), this provides the foundation for a surprisingly cogent theory of the infinite and infinitesimal. Leibniz had already recognized in 1674 (as can be seen by the above quotation) that the infinite can be treated as a *fiction*. This signifies that it can be treated *as if* it is an entity, in a certain respect, provided that the statements in which it occurs can be interpreted without supposing there is such a thing. For instance, in an infinite series, the infinite multiplicity of terms can be treated as if they are a collection of terms added together, provided a workable account of this infinite 'sum' can be given which does not presume this. In 1676, Leibniz finds a definition of the sum of a converging infinite series in keeping with his syncategorematic account of the infinite:

> Whenever it is said that a certain infinite series of numbers has a sum, I am of the opinion that all that is being said is that any finite series with the same rule has a sum, and that the error always diminishes as the series increases, so that it becomes as small as we would like. For numbers do not *in themselves* go absolutely to infinity, since then there would be a greatest number. (A VI 3, 503; LoC 98–99)

This is a good example of how Leibniz's philosophy of the actual infinite is supposed to work: you can still do mathematics with infinite quantities. Under certain conditions, they can be treated as fictional wholes, in the same way that the sum of this infinite series is a fictional sum, and the justification is in terms that, after Cauchy, we would now express in terms of ε and δ. Leibniz's definition here of the sum of a converging infinite series is equivalent to the modern one in terms of a limit of partial sums, and does not involve first taking an actual infinity of terms and then forming a sum of them.[16]

---

[16] As this paper goes to print, my attention has been drawn to a paper by David Rabouin (2011), who gives a reading entirely compatible with the present one by comparing Leibniz's philosophy of the infinite with that of Nicholas of Cusa. See also Ishiguro (1990), who was one of the first to argue that Leibniz can allow for the success of treating the infinite and infinitely small *as if* they are entities (under certain conditions), and that it is this that allows him to claim that mathematical practice is not affected by whether one takes them to be real or not. Philip Beeley (2009) gives a subtly different reading, interpreting such claims as instances of Leibniz's pragmatism. His paper is highly recommended for the intriguing connections he traces between infinity, conceptual analysis, the divine mind and the universal characteristic in the development of Leibniz's thought.

3   The Leibnizian Analysis of Matter

Turning now to the analysis of matter, we find the kind of picture that Cantor was proposing where continuous aetherial matter would be composed out of an (uncountable) infinity of substances ruled out in principle. The continuum, Leibniz claims, is something ideal, whereas what is real is an aggregate of unities. Here are some typical aphorisms to this effect:

> In actuals, simples are prior to aggregates, in ideals the whole is prior to the part. The neglect of this consideration has brought forth the labyrinth of the continuum. (To Des Bosses, 31st July 1709; GP II 379; Russell 245)
>
> Actuals are composed as is a number out of unities, ideals as a number out of fractions: the parts are actual in the real whole, not in the ideal whole. In fact we are confusing ideals with real substances when we seek actual parts in the order of possibles, and indeterminate parts in the aggregate of actuals, and we entangle ourselves in the labyrinth of the continuum and inexplicable contradictions. (To De Volder, 19/1/1706: GP II 282)
>
> It is the confusion of the ideal and the actual that has embroiled everything and produced the labyrinth *of the composition of the continuum.* Those who compose a line from points have quite improperly sought first elements in ideal things or relations; and those who have found that relations such as number and space (which comprise the order or relation of possible coexistent things) cannot be formed from an assemblage of points, have for the most part been mistaken in denying that substantial realities have first elements, as if there were no primitive unities in them, or as if there were no simple substances. (Remark on Foucher's Objections (1695); GP IV 491)

On the basis of such claims, Nicholas Rescher has concluded that the solution Leibniz is offering to the continuum problem is that in the mathematical continuum the whole is prior to the parts, but in the metaphysical one, the parts (monads) are prior to the whole. For it is not case that "both the indivisible constituent and the continuum to which it belongs [can] both at once be real":

> In mathematics the continuum, the line, is real and the point is merely the ideal limit of an infinite subdivision. In metaphysics only the ultimate constituents, the monads, are actual, and any continuum to which they give rise is but phenomenal. This is the Leibnizian solution of the paradoxes of the continuum. (Rescher 1967, p. 111)

Thus Rescher attributes to Leibniz a two-tiered ontology: the metaphysical, in which the actual monads are constituents of a phenomenal continuum, and the mathematical, in which the line segments are real and points are their merely ideal limits.

There are many problems with this analysis, however. Regarding the mathematical, Leibniz is clear that all mathematical objects are *ideal entities*, and (after 1676) that a point is always an *endpoint* of a line segment, never the ideal limit of an infinite subdivision. The infinitesimals of his mature theory, moreover, are *not* indivisible, and are in any case fictions rather than actual parts. And conversely, on the metaphysical side, any phenomenal whole resulting from an aggregate of monads ("secondary matter" or "body") must be well founded or *real,* and cannot therefore have the indeterminate parts characteristic of the continuous: "But in real things, namely bodies, the parts are not indefinite, as they are in space, a mental thing" (To

De Volder, 6/30/1704: GP II 268); "in actuals there is nothing indefinite—indeed in them every division that can be made is made... the parts are actual in the real whole..." (To De Volder, 1st January 1706: GP II 282). Consequently such a real phenomenal whole is *discrete* rather than a continuum: "Matter is not continuous, but discrete... [It is the same with changes, which are not really continuous.]" (To De Volder, 11th October 1705; GP II 278).[17]

Many of these criticisms were made by J. E. McGuire in his (1976), who consequently ascribed to Leibniz a three-tiered ontology, consisting in the actual, the phenomenal and the ideal.[18] In distinguishing these levels he makes use of the distinction in Leibniz between division and resolution, a distinction of crucial importance in understanding Leibniz's views, as we shall see in due course. Thus in Leibniz's mature metaphysics, space and time are characterized as ideal, or "entities of reason" (*entia rationis*) (McGuire 1976, p. 307; Hartz and Cover 1988, 504, 513). As continuous entities, they are arbitrarily divisible, though not composed of parts (McGuire 309; Hartz and Cover 505). They are *resolvable* into points or instants, but neither divisible into nor composed out of these, since points and instants are mere modalities (McGuire 1976, pp. 309–310). Well-founded phenomena, on the other hand are extended aggregates, and as such presuppose a plurality of entities from which their extension results. Thus they are *resolvable* into units of substance. They are also *divisible* into actual parts, and *composed* of these parts. McGuire concludes, however, that the only *actuals* are the substantial unities into which phenomena are resolved. Being simple, these actual substances themselves "can be neither composable, nor resolvable, nor divisible" (310).

The difficulty with this last claim is as follows. If monads or simple substances are the only actuals, then they must be the actual parts from which phenomena are composed, as McGuire duly concludes: "the 'actual parts' of extended things are non-extended substances" (306). But this flies in the face of what Leibniz says, as McGuire is perhaps tacitly acknowledging by his use of scare quotes.[19] Monads are supposed to be simples, entities into which bodies and motions are *resolved*, not parts out of which they are *composed*. As Leibniz writes to Burcher de Volder in 1704 in a typical passage,

> But, accurately speaking, matter is not composed of constitutive unities, but results from them, since matter or extended mass is nothing but a phenomenon founded in things, like a

---

[17] On an Aristotelian understanding of the continuous, a continuum is unbroken, and has no actual boundaries within. A line that is actually divided into contiguous line segments is therefore no longer regarded as continuous but as possessing discrete parts, notwithstanding the fact that there are no gaps between these contiguous segments. Thus when Leibniz describes matter as "discrete" he means actually divided into contiguous parts, but as still forming a plenum. For an engaging and informative history of this conception of the continuum as cohesive and unbroken, from Aristotle to present-day smooth infinitesimal analysis, see Bell 2006.

[18] Here he has been followed by Glenn Hartz and Jan Cover (1988), who contend that Leibniz changed his position from a 2-realm view to the 3-realm view of his mature metaphysics, after a period of transition between the years 1695 and 1709.

[19] Although Hartz and Cover criticize McGuire for his "misuse... of 'actual' to distinguish monads from bodies" (1988, p. 519), they nonetheless assert that "extension conceived as an abstract continuum has no actual parts, but extended bodies do have such parts: they are the genuine composites whose actual parts are Leibniz's 'atoms of substance' (cf. L 539, GP II 282)." (1988, p. 497).

> rainbow or mock-sun, and all reality belongs only to unities… Substantial unities, in fact, are not parts but foundations of phenomena. (To De Volder, 30th June, 1704: GP II 268)

The actual parts of phenomenal bodies, in fact, are not substances, but actually existing parts, as opposed to the indefinite parts into which a continuous body is divisible. Indeed, they are always mentioned by Leibniz in the context of an *actual division*. In the letter to De Volder of 19/1/1706, Leibniz writes "in actuals there is nothing indefinite—indeed, in them any division that can be made, is made" (GP II 282). And in his Remarks of 1695, the points marking the possible divisions of an abstract line are contrasted with the "the divisions actually made, which designate these points in an entirely different manner" (GP IV 49). Now since only phenomenal bodies can be divided (simple substances are indivisible), this means that Leibniz's "actual parts" must be parts of actually divided phenomenal bodies; and these parts will again be bodies. This is confirmed more explicitly in the following passages:

> But in real things, that is, bodies, the parts are not indefinite (as they are in space, a mental thing), but are actually assigned in a certain way, as nature actually institutes the divisions and subdivisions according to the varieties of motion, and… these divisions proceed to infinity… (to De Volder, June 30th, 1704: GP II 268).
> We should think of space as full of matter which is inherently fluid, capable of every sort of division, and indeed actually divided and subdivided to infinity; but with this difference, that how it is divisible and divided varies from place to place, because of variations in the extent to which the movements in it run the same way. (*New Essays*, Preface; Leibniz 1981, p. 59)

As these passages indicate, the actual divisions of matter are determined by the "varieties of motion". This is premised on the idea that any given body is individuated by its parts all having a motion in common, so that parts with differing motions will be actually divided from one another, as Descartes had in fact argued in his *Principles of Philosophy* (II, § 34–35). But in a plenum, according to Leibniz, every body is acted upon by those around it, causing differentiated motions in its interior, and thus dividing it. Since this is the case for every body, the division will proceed to infinity. Therefore matter is actually infinitely divided. Moreover, since what is divided is an aggregate of the parts into which it is divided, it will be an *infinite aggregate*. This argument is stated by Leibniz on numerous occasions throughout his *oeuvre*, including in the *Monadology* (1714): "every portion of matter is not only divisible to infinity, as the ancients realized, but is actually subdivided without end, every part into smaller parts, each one of which has its own motion." (WFPT 277). A particularly explicit example occurs in an unpublished fragment probably dating from 1678–1679:

> Created things are actually infinite. For any body whatever is actually divided into several parts, since any body whatever is acted upon by other bodies. And any part whatever of a body is a body by the very definition of body. So bodies are actually infinite, i.e. more bodies can be found than there are unities in any given number. (c. 1678–1679; A VI 4, 1393; LoC 235)

Significantly, the notion of the actual infinite Leibniz appeals to here is precisely the *syncategorematic* notion explained above: bodies are actually infinite in the sense that for any finite number, there are actually (not merely potentially) more bodies than this. Regarding the actual infinite, then, there is a perfect consilience between Leibniz's mathematics and his natural philosophy.

One consequence of this is that a body cannot be a true whole. For since every body is an infinite aggregate of its parts, it is an infinite whole; and, as we have seen, Leibniz held infinite wholes to be fictions. In an earlier work (Arthur 1989) I suggested that this explains in part why Leibniz held that bodies are phenomenal: since he regarded any substance as a true unity, bodies, being only fictional unities, would not qualify as substances. If a phenomenon is something that appears to the senses but is not a substance, then bodies, insofar as they really appear to the senses, must qualify as real phenomena. Thus the fact that bodies are phenomena is explained in part by Leibniz's doctrine of the actual infinite.

To this, two objections can be made. First, as Gregory Brown objected, a unity is not the same as a whole (Brown 2000, p. 41). Since Leibniz held that no substance can be composed of parts, no substance can be a whole, whether fictional or true.[20] Secondly, as Russell had already perceptively observed in 1900, although it is true that Leibniz identified bodies as infinite aggregates, and these as "corresponding to the phenomena", he also claimed that *all* aggregates are phenomenal, even finite ones: "A collection of substances does not really constitute a true substance. It is something resultant, which is given its final touch of unity by the soul's thought and perception." (Leibniz, *New Essays*, 1981, p. 226). Thus, as Russell remarks, "even a finite aggregate of monads is not a whole per se. The unity is mental or semi-mental" (Russell 1900, p. 116). This conclusion is a consequence of Leibniz's nominalism about aggregates, again accurately epitomised by Russell: "Whatever is real about an aggregate is *only* the reality of its constituents taken one at a time; the unity of a collection is what Leibniz calls semi-mental (GP II 304), and therefore the collection is phenomenal although its constituents are all real." (115)

When Russell was writing his book on Leibniz in 1900, he had still not encountered Cantor's set theory and was sympathetic to Leibniz's doctrine "that infinite aggregates have no number", describing it as "one of the best ways of escaping from the antinomy of infinite number" (117). But whatever reservations he may have harboured then about number in connection with infinite aggregates, he certainly had none about finite aggregates, and believed the doctrine of the phenomenality of aggregates to be a serious deficiency in Leibniz's philosophy. Russell took this doctrine to be a consequence of Leibniz's deriving his metaphysics from his logic, as "the assertion of a plurality of substances is not of this [subject-predicate] form—it does not assign predicates to a substance" (116).) But, he argued, if it is "the mind, and the mind only, [that] synthesizes the diversity of monads", then "a collection, as such, acquires only a precarious and derived reality from simultaneous perception" (116). So he confronted Leibniz with a dilemma:

---

[20] Here I owe a profound debt to my former student Adam Harmer, who first persuaded me of the significance of Leibniz's "mereological nihilism" for his notion of corporeal substance: this cannot be, as I had formerly supposed, a substance with a body that is a true whole at any given time.

> For the present it is enough to place a dilemma before Leibniz. If the plurality lies *only* in the percipient, there cannot be many percipients, and the whole doctrine of monads collapses. If the plurality lies *not* only in the percipient, then there is a proposition not reducible to the subject-predicate form, the basis for the use of substance has fallen through, and the assertion of infinite aggregates, with all its contradictions, becomes quite inevitable for Leibniz. The boasted solution of the difficulties of the continuum is thus resolved into smoke, and we are left with all the problems of matter unanswered. (Russell 1900, p. 117)

Even if we set aside Russell's mistaken belief that Leibniz derived his metaphysics of substance from a commitment to subject-predicate logic, though, there still remains a dilemma, given his reading of Leibniz's stance on plurality. For if plurality lies only in the percipient, then Leibniz is not entitled to assert that there is objectively more than one substance, and his system collapses into a monism; the infinite plurality of parts into which matter is divided must likewise exist only in the mind. Whereas if plurality is mind-independent, this seems to deprive Leibniz of any ground for denying the principle that to every aggregate there corresponds a number. In that case, Russell suggests, Leibniz would be forced to concede that there is infinite number, and he would fall into the very antinomies he was trying to avoid.

In fact, however, Russell's dilemma is based on a mistaken reading of Leibniz's doctrine of aggregates. It is not the plurality or aggregate that lies only in the percipient, but the perception of it *as a unity*. It is not the plurality itself that is contributed by perception, but the aggregate conceived as an entity *distinct from* its constituents. The *reality* of the aggregate does indeed consist in that of the constituents of the aggregate, just as Russell had described. What that means however is that if one has a flock of twenty sheep, say, each of these sheep exists independently of anyone perceiving it. If they are conceived or perceived together as making up a flock, then, according to Leibniz's doctrine, it is the flock *as distinct from its constituents* whose existence consists in those constituents being conceived or perceived together. This is consistent with Leibniz's position that numbers are ideal entities: the multiplicity of sheep exists independently of anyone numbering, but the numbering of the sheep as twenty requires someone to conceive them as making up a score or viguple, as a twenty. Thus the plurality itself is not mind-dependent, but only the judgement of it as forming a unity is, as is the applying to it of a number. In the same way, the divisions in matter are actual: they are not the result of any mental judgement, but of the internal motions of matter which are responsible for the divisions.

Leibniz is perfectly explicit on this point, as for instance in this passage from a letter to De Volder:

> I think that that which is extended has no unity except in the abstract, namely when we divert the mind from the internal motion of the parts by which each and every part of matter is, in turn, actually subdivided into different parts, something that plenitude does not prevent. Nor do the parts of matter differ only modally if they are divided by souls and entelechies, which always persist. (to De Volder, 3 April 1699; GP II 282)

Why, then, does Leibniz insist that matter is phenomenal? Again, his argument is laid out quite explicitly, both to Arnauld and to De Volder. It depends, as Russell recognized, on a "very bold use" (Russell 1900, p. 115) of his nominalist principle that *the reality of an aggregate derives only from the reality of its constituents*. As he explained to Arnauld, it follows from this that anything which, like matter, is a being by aggregation, must presuppose true unities from which it is aggregated:

> I believe that where there are only beings by aggregation, there will not in fact be any real beings; for any being by aggregation presupposes beings endowed with a true unity, because it derives its reality only from that of its constituents. It will therefore have no reality at all if each constituent being is still a being by aggregation, for whose reality we have to find some further basis, which in the same way, if we have to go on searching for it, we will never find. (to Arnauld, 30th April 1687; GP II 96; WFPT 123)

If a body is the aggregate of the parts into which it is divided, then its reality consists in the parts alone, and not in their being perceived as one. But since each of these parts is further divided, the argument iterates: the body is a perceived unity and a plurality of parts, but each of these parts is also a perceived unity and a plurality of parts, and so on down. If there are no true unities, then, given infinite division, the reality of body will elude analysis: it will reduce to a pure phenomenon. If, on the other hand, there exist true unities in the body, then the body's reality will reduce to the reality of these, while its unity will consist in their being perceived together. It will then be what Leibniz in his correspondence with De Volder calls a "quasi-substance", a plurality of substances with no substantial unity.[21] Leibniz repeats this argument to De Volder:

> Anything that can be divided into many (already actually existing) things is aggregated from many things, and a thing that is aggregated from many things is not one except in the mind, and has no reality except that which is borrowed from what it contains. From this I then inferred that there are therefore indivisible unities in things, because otherwise there will be no true unity in things and no reality that is not borrowed, which is absurd. For where there is no true unity then there is no true multitude. And where there is no reality except that which is borrowed, there will never be reality, since this must in the end be proper to some subject. (30 June 1704; GP II 267)

Now it is important to appreciate that Leibniz does not identify the true unities that he claims must be in body with the various actual parts into which it is divided. Body is divided into parts, but resolved into unities. The unities are presupposed by the nature of a body as an aggregate: its reality must reduce to the reality of its constituents.[22] The actual parts, on the other hand, are the result of a motion in common that is actually instituted in matter. But this does not prevent there from being other motions within this part of matter: in fact, of course, Leibniz argues that there are always such differentiated motions in any part of matter, and this is what results in its being infinitely divided.[23] But this also means that no part of matter can be a

---

[21] See for instance Leibniz's letter to De Volder of 19 November 1703, in Lodge 2009, p. 445, 2013, p. 279.

[22] I have given a fuller analysis of Leibniz's notion of *presupposition* in Arthur 2011. If A *presupposes* B, then B *is in* A, and A *contains* B. These are equivalent technical notions of wide-ranging application, for which Leibniz gives a formal treatment. I argue that those things are *constituents* of A that are presupposed in every part of A and are not themselves further resolvable, such as points in a line, or simple substances in matter.

[23] Cf. Levey 1999, pp. 144–145: "Adjacent parcels of matter form a cohesive whole in virtue of their sharing a motion in common (*motus conspirans*), but this is consistent with each parcel hav-

true unity. Moreover, because all parts of matter are constantly jostling one another, the divisions of matter differ from one instant to another. This means that no part of matter remains the same—i.e. has exactly the same shape and size—through time. The true unities, on the other hand, are precisely things that remain the same through time.

What then is the connection between the true unities presupposed by the reality of body and the actual parts into which matter is divided? The answer, in short, is motion, and the foundation Leibniz provides for it. As we have seen, the parts of matter are actually distinguished from one another by their motions. Motion, on the other hand, is not fully real, according to Leibniz. Because of the relativity of motion, it is impossible to say to which of several bodies in relative motion it belongs. There must nevertheless be some subject of motion, or all motion will be a pure phenomenon. There must also be some foundation for the real distinction of the differing motions (more accurately, tendencies to motion) that exist at each instant, or else there will be no objective basis for distinguishing the actual parts of matter.[24]

This is where Leibniz's revamped notion of substance comes into play. The argument so far has been that there must be real unities in matter, and also that there must be some principles by which the differing motions in matter at any instant might be distinguished. These desiderata are both satisfied by Leibniz's conception of the unities or substances as beings capable of action, for which it is necessary for them to be repositories of *force*. On the one hand, force is "an attribute from which change follows, whose subject is substance itself" (to De Volder, 3 April 1699; GP II 170; Lodge 2009, p. 313, 2013, p. 73); on the other, it involves an endeavour or striving (*nisus*), and this is what the reality of motion consists in: "there is nothing real in motion but the momentary state which a force endowed with a striving for change must produce" (*Specimen Dynamicum*, 1695; GM VI 236; WFPT 155). This force is thus the foundation for the motion of any actual part of matter at any instant. It is an *entelechy* in the sense that it remains self-identical through the changes of state that it brings into actuality: it is the real foundation at any instant for the

---

ing a motion of its own that divides and distinguishes it from the others. Also there can be further differing motions within each parcel that distinguish *its* parts." Thus having a motion in common is sufficient to individuate a raindrop, for instance, but does not preclude there being a variety of motions within the drop which, according to the Cartesian criterion, divide it within. Just as a line segment can be divided into further line segments, and these again, without limit, it is not necessary that an infinite division should issue in points or infinitesimals, contra Gregory Brown's assertion: "For that the divisions within matter must finally resolve themselves into infinitesimals or minima is something that seems to be guaranteed by Leibniz's assumption that *every* part of matter is divided to infinity" (Brown 2000, p. 34). Levey identifies this "folds" model as a third model for Leibnizian infinite division, in addition to his "divided block" and "diminishing pennies", although he later describes both it and the divided block models as "incoherent, given Leibniz's metaphysics of matter" (Levey 1999, p. 148).

[24] "For at the present moment of its motion, not only is body in a place commensurate with itself, but it also has an endeavour or striving to change place, so that from its subsequent state follows per se from its present one by the force of nature. Otherwise, ... there would be absolutely no distinction between bodies, seeing as in a plenum of mass that is uniform in itself the only means for distinguishing them is with respect to motion" (*On Nature Itself*, GP IV 513; WFPT pp. 218–219).

motion individuating the actual part of matter that is its body. The differing tendencies to motion to which the entelechies in matter give rise are what make actual the various parts into which matter is divided at different instants.

Leibniz does not much stress the role of his entelechies in making the parts of matter actual, but it is there if you look for it. We have already quoted above his riposte to De Volder's claim that the parts of mattercould be distinguished only modally: "Nor do the parts of matter differ only modally if they are divided by souls and entelechies, which always persist." (GP II 282) There is also this passage:

> Since, therefore, primitive entelechies are dispersed everywhere throughout matter—which can easily be shown from the fact that principles of motion are dispersed throughout matter—the consequence is that souls also are dispersed everywhere throughout matter. (GP VII 329; Russell 1900, p. 258)

Thus matter is actually divided by its different motions; each of these presupposes a principle, that is, a substance that is the subject of the changes, and a force that results in the changes occurring. Because each part of matter is further divided, there are such substances or true unities (Greek: *monada*) everywhere:

> If there were no divisions of matter in nature, there would not be any diverse things, or rather there would be nothing but the mere possibility of things: but the actual division in masses makes distinct the things that appear, and presupposes (*supponit*) simple substances. (unsent draft to De Volder, 1704-5; GP II 276)
> Since monads or principles of substantial unity are everywhere in matter, it follows from this that there is also an actual infinity, since there is no part, or part of a part, which does not contain monads. (to Des Bosses, 14 Feb 1706; GP II 301; LDB 25; Russell 1900, p. 129)

This, then, is Leibniz's argument for the actually infinite plurality of monads or simple substances. Because monads are presupposed in every actual part of matter, and matter is infinitely divided, there are actually infinitely many monads. Moreover, this actual infinite is understood syncategorematically, in perfect agreement with his mathematics of the infinite: their multiplicity is greater than any given number:

> In actuals, there is nothing but discrete quantity, namely the multiplicity of monads or simple substances, which is greater than any number whatever in any aggregate whatever that corresponds to the phenomena. (to De Volder, 19th January, 1706; GP II 282)

Thus we have the following contrast between finite aggregates and infinite ones. A finite aggregate is the whole formed by its parts, by analogy with the formation of the natural numbers from unities: $1+1+1=3$. Addition, however, is a mental operation: numbers as such are ideal, as are flocks interpreted as entities distinct from their members. One can imagine an infinite aggregate as similarly corresponding to an infinite sum, as the whole formed by an infinite addition: $1+1+1+\ldots=\infty$. But the idea of an infinite whole leads to contradiction, since then a part will be equal to the whole. Thus it is a fiction: while one can work with infinite sums of infinitely small elements under certain well defined conditions without falling into error, there is no such thing in actuality as an infinite sum or an infinite addition: the infinite is not a true whole, and there is no such thing as an infinite number.

In conclusion: according to Leibniz, to say that there are actually infinitely many parts of matter at each instant is to say that there are so many that for any finite number one assigns, there are more. But each of these actual parts presupposes true unities. These constitute what is real about the bodies, since the reality of the aggregate reduces to the reality of its constituents, while the unity of the aggregate is supplied by a perceiving mind. Therefore, "since there is no part, or part of a part, which does not contain monads", bodies are infinite aggregates of monads: in any body there are more monads than can be assigned. For Leibniz, there are actually infinite aggregates, but—in contrast to Cantor—there are no infinite numbers.

Acknowledgements I am indebted to Michael Detlefsen, Sam Levey, David Rabouin, and Ohad Nachtomy for helpful criticisms and comments on a previous version of this paper, and to Paul Lodge for making available a preliminary draft (Lodge 2009) of his translation volume of the Leibniz-De Volder Correspondence, as well as to an anonymous referee for helpful suggestions.

References

Arthur, R. T. W. 1989. Russell's conundrum: On the relation of Leibniz's monads to the continuum. In *An intimate relation: Studies in the history and philosophy of science,* eds. James Robert Brown and Jürgen Mittelstraß, 171–201. Dordrecht: Kluwer Academic Publishers.

Arthur, R. T. W. 2001. Leibniz on infinite number, infinite wholes and the whole world: A reply to Gregory Brown. *Leibniz Review* 11:103–116.

Arthur, R. T. W. 2008. Leery Bedfellows: Newton and Leibniz on the status of infinitesimals. In *Infinitesimal differences: Controversies between Leibniz and his contemporaries,* eds. Ursula Goldenbaum and Douglas Jesseph, 7–30. Berlin: de Gruyter.

Arthur, R. T. W. 2011. Presupposition, aggregation, and Leibniz's argument for a plurality of substances. *The Leibniz Review* 21:91–115.

Bassler, O. Bradley. 1998. Leibniz on the indefinite as infinite. *Review of Metaphysics* 51:849–874.

Beeley, Philip. 2009. Approaching infinity: philosophical consequences of Leibniz's mathematical investigations in Paris and thereafter. In *Philosophy of the young Leibniz,* eds. Mark Kulstad, Mogens Lærke, and David Snyder, 29–47. Suttgart: Franz Steiner.

Bell, John L. 2006. *The continuous and the infinitesimal in mathematics and philosophy*. Milan: Polimetrica.

Benardete, Jose. 1964. *Infinity: An essay in metaphysics*. Oxford: Oxford University Press.

Breger, Herbert. 2008. Natural numbers and infinite cardinal numbers. In *Kosmos und Zahl : Beiträge zur Mathematik- und Astronomiegeschichte, zu Alexander von Humboldt und Leibniz,* ed. Hecht Hartmut, et al., 309–318. Stuttgart: Steiner.

Brown, Gregory. 1998. Who's afraid of infinite numbers? Leibniz and the world Soul. *Leibniz Review* 8:113–125.

Brown, Gregory. 2000. Leibniz on wholes, unities and infinite number. *Leibniz Review* 10:21–51.

Brown, Gregory. 2005. Leibniz's mathematical argument against a soul of the world. *British Journal for the History of Philosophy* 13 (3): 449–488.

Cantor, Georg. 1932. *Gesammelte Abhandlungen,* ed. Ernst Zermelo. Berlin: 1932; repr. Hildesheim: Georg Olms, 1962).

Dauben, J. 1979. *Georg Cantor: His mathematics and philosophy of the infinite*. Boston: Harvard University Press.

Euler, Leonhardt. 1843. *Letters of Euler on different subjects in natural philosophy addressed to a German princess, David Brewster*, (S.). vol. 2. New York: Harper and Brothers.

Gerhardt, C. I., ed. 1849–63. *Leibnizens Mathematische Schriften*. Berlin and Halle: Asher and Schmidt. (reprint ed. Hildesheim: Georg Olms, 1971. 7 vols; cited by volume and page, e.g. GM II 157).

Gerhardt, C. I., ed. 1875–90. *Die Philosophische Schriften von Gottfried Wilhelm Leibniz*. Berlin: Weidmann. (reprint ed. Hildesheim: Olms, 1960), 7 vols; cited by volume and page, e.g. GP II 268).

Hartz Glenn, A., and Cover A. Jan. 1988. Space and time in the Leibnizian metaphysic. Nous 22:493–519.

Hettche, Matt. 2006. Christian Wolff. Stanford encyclopedia of philosophy. http://plato.stanford.edu/entries/wolff-christian/. Accessed 12 July 2012.

Ishiguro, Hidé. 1990. Leibniz's philosophy of logic and language. 2nd ed. Cambridge: Cambridge University Press.

Leibniz, G. W. 1923- *Sämtliche Schriften und Briefe, ed. Akademie der Wissenschaften der DDR*. Darmstadt: Akademie-Verlag (cited by series, volume and page, e.g. A VI 2, 229).

Leibniz, G. W. 1981. *New essays on human understanding*. Transl. & ed. Peter Remnant and Jonathan Bennett. Cambridge: Cambridge University Press.

Leibniz, G. W. 2001. *The Labyrinth of the continuum: Writings on the continuum problem* [1672–1686. Ed., sel. & transl. R. T. W. Arthur]. New Haven: Yale University Press; abbreviated LoC with page number.

Levey, Samuel. 1999. Leibniz's constructivism and infinitely folded matter. In *New essays on the rationalists,* eds. Rocco Gennaro and Charles Huenemann, 134–162. New York: Oxford University Press.

Levey, Samuel. 2008. Archimedes, infinitesimals, and the law of continuity: On Leibniz's fictionalism. In *Infinitesimal differences: Controversies between Leibniz and his contemporaries,* eds. Ursula Goldenbaum and Douglas Jesseph, 107–133. Berlin: de Gruyter.

Lodge, Paul. 2009. The Leibniz-De Volder correspondence. Preliminary draft of (Lodge 2013); http://users.ox.ac.uk/~mans1095/devolder.htm. Accessed 12 July 2012.

Lodge, Paul. 2013. *The Leibniz-De Volder Correspondence* (Transl., ed. and with an introduction by Paul Lodge). New Haven: Yale University Press.

Mancosu, Paolo. 2009. Measuring the size of infinite collections of natural numbers: Was cantor's theory of infinite number inevitable? *The Review of Symbolic Logic* 2 (4): 612–646.

McGuire, J. E. 1976. 'Labyrinthus Continui': Leibniz on substance, activity and matter. In *Motion and time, space and matter,* eds. P. K. Machamer and R. G.,Turnbull, 290–326. Columbus: Ohio State University Press.

Rabouin, David. 2011. Infini mathématique et infini métaphysique: D'un bon usage de Leibniz pour lire Cues (et d'autres). *Revue de métaphysique et de morale* 70 (2): 203–220.

Rescher, Nicholas. 1967. *The philosophy of Leibniz*. Englewood Cliffs: Prentice-Hall.

Russell, Bertrand. 1900. *A critical exposition of the philosophy of Leibniz*. 2nd ed. Cambridge: Cambridge University Press (1937, reprinted London, Routledge, 1992).

Russell, Bertrand. 1919. Introduction to mathematical philosophy. London: George Allen and Unwin.

van Atten, Mark. 2011. A note on Leibniz's argument against infinite wholes. *British Journal for the History of Philosophy* 19 (1): 121–129.

Wolff, Christian. 1737. Cosmologia generalis. Frankfurt: Olms (1964. reprint Hildesheim).

Woolhouse, R. S., and Richard Francks, eds. 1998. *G. W. Leibniz: Philosophical texts*. Oxford: Oxford University Press (cited as WFPT 101, etc.).

Comparability of Infinities and Infinite Multitude in Galileo and Leibniz

Samuel Levey

1 Introduction

Galileo's discussion of the infinite in *Discourses and Mathematical Demonstrations Concerning Two New Sciences* (1638) hardly wants for recognition. But its importance for Leibniz's philosophy has not always been appreciated. Nor, I think, has Galileo's own view of the infinite in *Two New Sciences* yet been properly understood. A close study of Galileo's paradox of the natural numbers and his answer to it can throw new light on Galileo's own position and, with its elements in view, the influence of Galileo on Leibniz comes into high relief. A number of new points of interpretation of Galileo will be on offer in what follows, some likely to be controversial. Contrary to the customary account, for instance, I hold that Galileo allows for judgments of equality among infinite classes; indeed they are readily found in his mathematical and philosophical work. As I see it, his celebrated denial that the terms 'greater', 'less' and 'equal', apply in the infinite is in fact limited to unbounded magnitudes, but consistent with judgments of cardinal equality among infinite multitudes that are bounded in magnitude and thus, as magnitudes, finite. Galileo's denial of comparability nonetheless poses a threat to two important mathematical principles, Euclid's Axiom and the Bijection Principle of Cardinal Equality, and I consider two sorts of strategies for reconciling those principles with Galileo's position. One strategy, suggested by Eberhard Knobloch,[1] appeals to Galileo's use of the distinction between *quanti* and *non quanti*. The other is due to Leibniz and involves a distinction between totalities and pluralities. I argue that the first strategy cannot save Galileo's account from having to relinquish at least one of the two mathematical principles. Leibniz's strategy offers a more promising way to escape from the paradox while leaving both principles intact, although it imposes a peculiar metaphysical cost of its own. Spelling out the details of Leibniz's solution further reveals

---

[1] Knobloch (1999; 2011).

S. Levey (✉)
Philosophy Department, Dartmouth College, Hanover, NH, USA
e-mail: samuel.s.levey@dartmouth.edu

© Springer Netherlands 2015
N. B. Goethe et al. (eds.), *G.W. Leibniz, Interrelations between Mathematics and Philosophy*, Archimedes 41, DOI 10.1007/978-94-017-9664-4_8

how intimately related his account of the term 'infinite' is to Galileo's discussion and draws out key contrasts between their respective views of comparability and their definitions of 'infinite'.

1.1   Galileo's Paradox of the Natural Numbers

Early in the discussion of the dialogue's First Day, Galileo offers a striking proof for the claim that "one infinity cannot be said to be greater or less than or equal to another" (EN 8:78/D40).[2] The context is one in which Galileo, via his spokesman Salviati, is looking to defend the coherence of the idea that a finite quantity such as a line or a solid might contain an infinity of indivisible points. Simplicio has detailed an objection: it seems a longer line would then contain an infinity of points greater than the infinity contained in a shorter line, implying an infinity greater than the infinite, "a concept not to be understood in any sense" (EN 8:77/D 39). Galileo's proof would cut off the objection by disallowing any comparison of size among 'infinites.' It is the proof itself, though, not the picture of matter being defended, that is our present concern.

Galileo takes the natural numbers as his example of an infinite and argues as follows. Since the natural numbers include both the square numbers and non-square numbers, there are more [esser più che] naturals than squares. Yet there are just as many squares as there are roots, since every root has its own square and every square its own root; and there are just as many naturals as roots, since every natural is a root and every root is a natural. So it follows that there are just as many [siano quanti] squares as naturals. We thus appear to have contradictory results: the natural numbers are both greater than and equal to the square numbers, which is absurd. (Cf. EN 8:78–79.)

Galileo's paradox of the natural numbers, then, appears to derive a contradiction from the idea that one infinite can be said to be greater or less than or equal to another. As is readily noted, the proof trades on two different standards for comparison. By the standard of 'proper inclusion', there are more naturals than squares since the natural numbers properly include the squares, i.e. the naturals include non-square

---

[2] Primary texts are abbreviated as follows. For Galileo: EN=*Opere*, Edizione Nazionale, ed. Antonio Favaro (Florence 1898). For Leibniz: A=Berlin Academy Edition, *Sämtliche Schriften un Briefe. Philosophische Schriften*. Series VI. Vols. 1–4. (Berlin: Akademie-Verlag, 1923–99); GP=Gerhardt, *Die Philosophischen Schriften*, Vols. 1–7. Ed. C.I. Gerhardt (Berlin: Weidmannsche, Buchhandlung 1875–1890); GM=*Mathematische Schriften von Gottfried Wilhelm Leibniz*, Vols. 1–7. Ed. C.I. Gerhardt (Berlin: A. Asher; Halle: H.W. Schmidt 1849–1863). References to EN, GP and GM are to volume and page numbers; those to A are to series, volume and page. Translations of Galileo generally follow those of Stillman Drake (abbreviated 'D'), *Galileo Galilei: Two New Sciences, Including Centers of Gravity and force of Percussion*, 2nd Ed, (Toronto: Wall and Emerson, Inc. 1974), and those of Leibniz generally follow Richard Arthur (abbreviated 'Ar'), *G.W. Leibniz: The Labyrinth of the Continuum: Writings on the Continuum Problem, 1672–1686* (New Haven: Yale University Press 2001). I have sometimes modified translations without comment.

numbers as well as all the square numbers. Or as we might say, the squares form a proper subclass of the natural numbers.[3] By the standard of 'one-one maps', however, there are just as many squares as naturals, since the two classes can be mapped one-one into each other, implying a 'one-one correspondence' (or *bijection*) between them. In the case of finite classes the standards are always in agreement: no finite class can be mapped one-one into one of its own proper subclasses, and finite classes are always greater than their proper subclasses. Only in the infinite can the standards conflict .

Galileo in effect treats the two standards of comparison as equally sound and suggests that we are mistaken to extend either one from the finite case to the infinite case. He recommends abandoning comparisons altogether in the infinite. History has instead taken sides in order to resolve the paradox, and it has favored the standard of one-one maps over that of proper inclusion. Classes X and Y are equal in size just in case there is a one-one correspondence between them, the proper inclusion of one in the other notwithstanding. Developments in transfinite set theory due to Cantor would establish this as a consistent approach to the paradox,[4] and subsequent orthodoxy was to hold, in Russell's words, "it is actually the case that the number of square (finite) numbers is the same as the number of (finite) numbers."[5]

That is all familiar enough. Writers on the topic tend to be orthodox Russellians on this point today. Galileo's own analysis of the paradox is hardly refuted by the preference of history,[6] however, and what he has to say is quite interesting. Here in the words of Salviati:

> I don't see how any other decision can be reached than to say that all the numbers [*tutti i numeri*] are infinitely many [*infiniti*]; all the squares infinitely many; all their roots infinitely many; that the multitude [*moltitudine*] of squares is not less than that of all numbers, nor is the latter greater than the former. And in the final conclusion, the attributes of equal, greater and less have no place in infinite, but only in bounded quantity [*quantità terminate*]. (EN 8:79/D 41)

The denial that such comparisons are possible in the infinite is Galileo's signature conclusion here. But his words convey a few more ideas worth drawing out. The

---

[3] A quick note on terminology. In what follows I sometimes use the terms 'class', 'subclass', etc., for convenience, but without meaning to imply that the many elements of a class thereby form a *set* or *totality* or other 'single object.' (Mostly here it will cause no harm to read 'class' and 'set' as equivalent, but sometimes it will lead astray, so beware.) Occurrences of 'class', etc., can, with appropriate shifts in syntax, always be replaced by suitable plural expressions—e.g., 'the natural numbers' instead of 'the class of natural numbers'—or by terms such as 'multitude' or 'plurality' that cancel the implication of one thing formed from many. In contexts in which greater precision is required to convey the intended meaning, and avoid unwanted implications, I shall use unambiguous terms.

[4] Cantor writes, "There is no contradiction when, as often happens with infinite aggregates, two aggregates of which one is a part of the other have the same cardinal number" (Cantor 1915, p. 75; noted in Parker (2009)).

[5] Russell (1913, p. 198).

[6] For a very illuminating discussion of the history of the idea of measuring the size of the natural number collections, as it has evolved up to Cantor, plus some contemporary alternatives to Cantor, see Mancosu (2009).

multitude of (natural) numbers is infinite, as are those of the squares and their roots. This is justified, presumably, by the one-one correspondences between the squares and the roots and the roots and the naturals. Thus it seems that even if a one-one correspondence isn't sufficient for claiming that two classes are equal, it is sufficient for claiming that a class is infinite if there is a one-one map from some infinite class into it—or, at least, a class is infinite if there is a one-one map from the natural numbers into it.

Galileo does not define 'infinite' explicitly in *Two New Sciences*. Still, his suggestion that a class is infinite if the natural numbers can be mapped into it can itself serve well as an intuitive definition. And his discovery that the class of natural numbers can be mapped one-one into a proper subclass of itself suggests a structural property of classes that would later be elevated by Dedekind into a definition of 'infinite': infinite classes are exactly those that can be mapped into one of their own proper subclasses. In Dedekind's words, 'A system S is said to be infinite when it is similar to a proper part of itself.'[7] In the terms of modern-day mathematics, Galileo discovered that the natural numbers are 'Dedekind infinite'.

1.2   Parts, Wholes and Euclid's Axiom

Unlike Dedekind, Galileo does not use the word 'part' or 'proper part' in an explicitly formal statement of his mathematical principles. The language of parts is not absent from his discussion, however, for he does say, in passing, that the squares form a part of the natural numbers—a 'tenth part' of the first hundred numbers, a 'hundredth part' of the first ten thousand, etc.—and that the non-squares form a 'greater part' [*maggior parte*] than the squares (EN 8: 79). But it is at best equivocal evidence that he means to use the language of parts and wholes for his technical mathematical vocabulary. And Galileo does not say that the natural numbers form a 'whole' or 'totality' or even a 'system'. His phrase *tutti i numeri* might suggest this, since *tutti* can have the force of 'whole' in some uses,[8] but taken straightforwardly what Galileo says is simply 'all the numbers' and likewise 'all the squares' and 'all the roots'. Moreover, his use of the term *moltitudine*, or 'multitude', in claiming that the multitude of squares, that of natural numbers and that of roots are all infinite, seems gauged to avoid the supposition that there is a single totality, a single mathematical object, made up of all the squares or all the naturals or all the roots.

The language of parts and wholes is nonetheless a natural one in which to frame the discussion, and it also offers an idea that likely lies behind Galileo's appeal to the standard of proper inclusion in his initial claim that the natural numbers as a

---

[7] In the usual formula: S is infinite if and only if there is a one-one map Φ from S into S with some element of S not in the range of Φ. *Was sind und was sollen die Zahlen?*, Sect. 64; cf. Sect 66.

[8] Crew and de Salvio's 1914 translation renders *tutti i numeri* as "the totality of all numbers" (cf. p. 31). In the same lines it also inserts 'number' in "the number of squares is infinite" and "the number of roots is infinite", where the corresponding term does not occur in the original. Drake's translation steers clear of those interpolations.

class are greater than the squares. That idea is the following principle: The whole is greater than the part. It is sometimes called 'Euclid's Axiom' for its occurrence as Common Notion V at the start of Book One of the *Elements*. If the square numbers form a part of the natural numbers, then by Euclid's Axiom the whole of the natural numbers must be greater than the part formed by the square numbers alone. When Dedekind says that in an infinite system the proper part is equal ('similar') to the whole, his position implies the falsity of Euclid's Axiom.

It is not hard to think Galileo's position has the same consequence, if somewhat more subtly. When he recommends that we drop the terms 'greater', 'equal' and 'less' from use in application to infinites, he is in effect abandoning Euclid's Axiom, at least in the case of the infinite. For the result would be to deny that the whole is greater than the part in this case, even if the equality of part and whole is not asserted. As we shall see, Leibniz interprets Galileo in just this way. At the moment it is enough to observe the potential implication for Euclid's Axiom, and to note that Galileo himself may not quite be committed to it, since it is not evident to what extent he embraces the language of parts and wholes for classes of numbers.

The minimal reading of Galileo's own position is just that the extension of Euclid's Axiom from the finite to the infinite is incorrect. He is explicit in warning against taking such extensions for granted. "These are some of those difficulties that arise", he writes, "that derive from reasoning about infinites with our finite minds and giving to them those attributes that we give to the finite and the bounded" (EN 8: 77–78/D39–40). It will then remain to say *why* it is incorrect to apply terms of comparison, or principles like Euclid's Axiom, in the infinite case. The paradox only points up an inconsistency, perhaps showing *that* the extension is invalid; it does not explain the underlying problem.

It is open to Galileo to deny the applicability of Euclid's Axiom in the infinite case without thereby rejecting the axiom itself, if a condition of its terms can be seen not to hold in the infinite. The involvement of the concepts of part and whole in the axiom indicates one possible avenue for doing this: if we should say that infinite multitudes cannot form wholes, then Euclid's Axiom will be seen not to apply in the infinite case without thereby being overturned by a counterexample. If there are no infinite wholes, then there is no whole that fails to be greater than its parts. Another possible avenue might be to identify some less explicitly stated condition of Euclid's Axiom, say, some requirement that the terms it compares be in some way 'measurable' or 'quantifiable', and then see if it can be denied that 'infinites' meet this condition. If it can be held that infinites are not measurable or quantifiable in the relevant way, then denying that Euclid's Axiom applies to them might fall short of rejecting the axiom itself. Again, its inapplicability would not be due to the existence of counterexamples.

I belabor those points because each of those two avenues has been suggested as a possible route of escape from the prospect of having to deny Euclid's Axiom. The first—denying that infinite multitudes form a whole—is proposed by Leibniz, who, commenting on Galileo's paradox of the natural numbers in his notes on *Two New Sciences*, writes:

> Hence it follows either that in the infinite the whole is not greater than the part, which is the
> opinion of Galileo and Gregory of St. Vincent, and which I cannot accept; or that infinity
> itself is nothing, i.e. that it is not one [*Unum*] and not a whole [*totum*]. (A VI, 3, 158/Ar 9)

As before, it is not quite true that Galileo says the whole is not greater than the part
in the infinite, since his mention of parts is not clearly committal and he does not
write of the number classes directly in terms of wholes. (Perhaps Leibniz, like some
of Galileo's later readers, sees the language of wholes [*totum*] in Galileo's phrase
tutti i numeri. This would not be surprising, since part-whole terminology was a
common feature of mathematical language in the seventeenth century.) We shall
consider Leibniz's own position in due course.

The second avenue of escape—that of denying that infinites are suitably measurable or quantifiable—is suggested as Galileo's own view by Eberhard Knobloch,
who calls attention to Galileo's careful distinction between those things which are
true quantities or 'quantified' (*quanti*) and those which are 'non-quantified' (*non-
quanti*) in the treatment of infinites and indivisibles in *Two New Sciences*. Knobloch
writes,

> An 'infinite quantity' ('quantità infinita') would according to Galileo's conception actually
> be a 'contradiction in terms', because an infinite lacks precisely those properties which
> characterize a quantity. [...]
> Correspondingly, the Euclidean axiom 'The whole is greater than the part' is not invalidated
> in the sense that the logical opposite is valid in the domain of infinite sets, that is, that an
> infinite set is smaller than or equal to one of its parts. Rather it is invalidated in the sense
> that it cannot be applied there, simply because there are not quantities which could be
> compared.[9]

Knobloch's analysis of Galileo's position is an important one, and we shall turn
shortly to consider the content of the distinction between *quanti* and *non-quanti*. For
now it is enough to note that it stands as an alternative to Leibniz's proposed route
of escape. Each one in its own way allows us to see how Galileo's suggestion that
infinites cannot be compared need not automatically imply the rejection of Euclid's
Axiom. And each would give us a way to explain why the axiom is not rendered
invalid: either there are no infinite wholes or there are no infinite quantities, and
hence there are no counterexamples.

1.3 The Same Question Revisited: The Bijection Principle

Just as Galileo's denial of comparison among infinites poses at least a *prima facie*
threat to Euclid's Axiom, so too it poses a threat to the idea that one-one correspondence between classes is a valid standard of equality—and, more generally, a threat
to the validity of using one-one maps to determine comparisons of size among sets
or classes. To the modern eye this may seem the more troubling element of Galileo's
position, since in the wake of Cantor, Euclid's Axiom has been set aside while the
standard of one-one maps has come into its own as a vital piece of mathematical
theory and practice.

---

[9] Knobloch (1999), p. 94.

A few words of clarification are in order about the intended principle of equality based on one-one maps and what we shall call it.[10] In contrast to some writers,[11] Galileo is addressing the idea of comparison between multitudes without any obvious presupposition of number.[12] His expressions for equality in the relevant passages are simply *tanti quanti* and *altrettanti* with the sense of 'precisely as many as'. Thus for now in representing his view we can step back from the idea of assigning a number to a multitude, and ask more minimally whether one-one correspondence, or bijection, implies equality of relative size, though of course the relevant notion of size is the cardinal one of 'many-ness' rather than, say, the metrical one of 'much-ness' or measure. Likewise for the related definitions of 'greater' and 'less': X is greater than Y if and only if Y can be mapped one-one into X but X cannot be mapped one-one into Y, and *vice versa* for 'less', but neither need be taken to imply a claim about number or absolute size. We shall adopt the precise if anachronistic label 'the Bijection Principle of Cardinal Equality'—or just 'the Bijection Principle'—for the principle that says X and Y are equal if and only if there is a one-one correspondence (bijection) between their elements. (Better still is the plural form: there are just as many Xs as Ys if and only if there is a one-correspondence between the Xs and the Ys.)

The issue before us is what to make of Galileo's abandonment of the Bijection Principle in the infinite. There are both conceptual and historical questions to consider. Take first the purely analytical question of whether this means that the Bijection Principle is simply invalid on Galileo's terms. Such a result did not follow in the case of Euclid's Axiom; there are ways of leaving the axiom intact while withholding it from the infinite. Yet unlike Euclid's Axiom, the Bijection Principle is not phrased in terms of parts and wholes. So if we follow Galileo in denying the comparability of infinites, there is no taking Leibniz's escape from the conclusion that infinites are counterexamples to the Bijection Principle by denying that infinites are wholes. If we take the Bijection Principle to apply only to *sets*, a clear version of Leibniz's tactic remains available. If an infinite multitude does not form a set—if, in

---

[10] My discussion is indebted to Parker (2009), who, defensibly, calls our two principles 'Euclid's Principle' and 'Hume's Principle'. If Galileo had not rejected the one-one maps standard in the infinite case, we should call it 'Galileo's Principle'. For reasons to think Archimedes made use of this principle in application to infinite classes, see Netz et al. (2001–2002).

[11] Notably those involved in discussion of a similar principle of equality sometimes called 'Hume's Principle': the number of Fs equals the number of Gs iff there is a one-one correspondence between the Fs and the Gs. The principle is so-called for Frege's reference, in Sect. 73 of the *Foundations of Mathematics*, to Hume's remark, in *Treatise* I.iii.1, "When two numbers are so combin'd as that one has always an unite answering to every unite of the other, we pronounce them equal." Yet both of those authors have their sights on slightly more restricted conditions than the ones Galileo considers. Frege takes one-one correspondence between classes to imply the existence of a *number* that measures them; Hume \t "*See* Principle" is expressly considering a standard of equality for numbers, where the numbers themselves are conceived as made up of units. It is in this vein that one-one correspondence is sometimes said to be a criterion of 'equinumerosity': equality of number.

[12] Or he appears to be doing so. Below I shall suggest his account of comparisons of infinite number classes turns out to involve an infinite number after all; that is, if, *per impossibile*, there were such a comparison, it would have to involve an infinite number.

Russell's phrase, it is a *proper class*—then perhaps it does not fall within the scope of the Bijection Principle, and so denying that it is comparable to other infinites via one-one maps does not thereby make it a counterexample to the principle. (Here our use of the neutral term 'class' rather than 'set' in referring to multitudes of things matters.[13]) Likewise, a version of Knobloch's strategy is available if we can read the Bijection Principle as tacitly requiring its terms of comparison to be *quanti* and then hold that infinite multitudes are *non-quanti*. The two strategies might be regarded as nearly equivalent, since a natural thought is that something is mathematically quantifiable or fit for mathematical measurement only if it can be understood as a 'single object,' such as a set. If infinites are not unities, but only uncollected multitudes like proper classes, they might on that ground be regarded as *non-quanti* and hence not candidates to be counterexamples to the Bijection Principle.

The question of whether Euclid's Axiom or the Bijection Principle face counterexamples matters because each has, in its time, been thought to capture or reflect something deep about the idea of a mathematical quantity. Euclid's Axiom was regarded as nearly constitutive of the very idea of quantity.[14] The Bijection Principle has something of the same position with respect to the idea of cardinality today. If either one were shown to be incorrect in clear cases, there would be reason to doubt whether the related understanding of quantity or cardinality were truly secure. Even damming up counterexamples on the far side of the distinction between the finite and the infinite is not automatically a satisfactory solution if we would otherwise take ourselves to see clearly that infinite classes meet the conditions of quantity or cardinality. If the properties we appeal to in the finite in order to justify our mathematical reasoning are patently also exemplified in the infinite but then lead into contradiction, we should doubt whether our original appeal was sound. That is, we should doubt whether it was sound unless we can explain why the extension of those properties to the infinite case is invalid.

Denying that infinites are wholes or that they are truly quantities can be a step in the direction of an explanation. "Our mathematical justifications in the finite case", we could say, "presupposed that the objects of study are wholes or quantities. The infinite case provided an initial appearance of this, but it was only an illusion. There simply are no 'infinite wholes' or 'infinite quantities' to which they may be applied. So there are no counterexamples to our principles." Of course, this only works if we have some ground for saying that infinites are not wholes or quantities independently of the paradox; otherwise, we are just left "wielding the big stick"—i.e., pointing to the contradiction—rather than offering an explanation.[15] I suspect the proposed escape routes, whether Leibniz's or the one Knobloch sees in Galileo,

---

[13] See footnote 3 above.

[14] Bolzano, for example, explicitly defended the primacy of Euclid's Axiom against the Bijection Principle, writing that even two sets that stand in a one-one correspondence "can still stand in a relation of inequality in the sense that the one is found to be a whole, and the other a part of that whole" (Bolzano 1950, p. 98). For discussion of Bolzano, see Parker (2009) and Mancoso (2009). Even Russell acknowledged that "the possibility that the whole and part may have the same number of terms is, it must be confessed, shocking to common sense" (1903, p. 358).

[15] On "wielding the big stick", see Michael Dummett (1994).

may turn out to be cases of the big stick and not truly explanations. But in any case it should be clear that there is something at stake here (if different stakes for audiences from different eras) in asking whether Euclid's Axiom or the Bijection Principle is in jeopardy of admitting counterexamples even in the infinite case.

1.4 Infinite Multitude and Non Quanti in Galileo

Back to Galileo and a few historical questions. Is Galileo's concept of *non quanti* meant to cover the case of the infinite multitude? I think the answer is no, or at least I believe that infinite multitudes do not automatically qualify as *non quanti* for Galileo, though in special cases they may do so. To cast enough light on Galileo's view here, a fairly close look at the texts will be required, though technical matters can be kept to a minimum.

The distinction between *quanti* and *non-quanti* in *Two New Sciences* occurs in connection with indivisibles, in particular with the idea that quantities such as lines or circles or solid bodies might contain or be resolved into infinitely many indivisible parts. Two sections of the dialogue are most explicit in discussing the idea of *non quanti*. In both, Galileo, in the voice of Salviati, appeals to the hypothesis of the composition of matter from indivisibles and the presence within it of indivisible vacua or void spaces to make sense of the possibility of the expansion or contraction in size of a finite quantity. These expansions and contractions ('rarefaction' and 'condensation') are themselves introduced to resolve the ancient paradox of the wheel, concerning the motion of concentric circles rolling along a line. The puzzle is that it appears that the smaller interior circle and the larger outer one will traverse equal distances in the course of a single revolution despite the difference in their circumferences. (See Fig. 1.)

Galileo approaches the problem by developing an analysis of the motion of concentric polygons and then extending it to that of the circles, taking the circles as

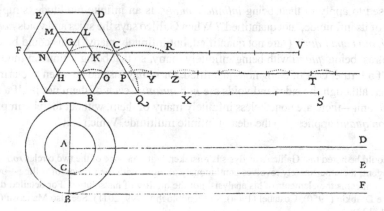

Fig. 1 Galileo's analysis of 'Aristotle's Wheel.' (EN 8: 68)

infinilateral polygons. The details of the analysis are, as Galileo's Sagredo notes, 'intricate,' and need not detain us.[16] The key point is that Galileo finds that the motion of the larger polygon passes over a line approximately equal in length to that traversed by the smaller polygon, but with the interposition of 'skipped over' void spaces into the line traversed by the smaller. (The presence of the void spaces compensates for the difference in the lengths directly marked out on the line through contact by the sides the two polygons, whose perimeters are, of course, unequal.) As the number of sides is increased, the sides and void spaces become smaller and the lengths of the lines measured out by the motions come closer to equality. Advancing now to the limit case of the motion of two concentric circles, the distinction between *quanti* and *non quanti* appears when Salviati says:

> And just so, I shall say, in the circles (which are polygons of infinitely many sides), the line passed over by the infinitely many sides of the larger circle, arranged continuously <in a straight line>, is equal in length to the line passed over by the infinitely many sides of the smaller, but in the latter case with the interposition of just as many voids [*d'altrettanti vacui*] between them. And just as the sides are not quantified [*lati non son quanti*], but are infinitely many [*ma bene infiniti*], so too the interposed voids are not quantified [*vacui non son quanti*], but are infinitely many; that is, for the former < line touched by the larger circle there are > infinitely many points all filled, and for the latter < line touched by the smaller circle>, infinitely many points, part of them filled points and part of them voids. (N 8:71/D 33)

As Knobloch observes,[17] Galileo shifts from having a little earlier spoken of the lines as 'measured' (*misurata*) by the finitely many sides of the finite polygons to saying only that they are 'passed over' (*passata*) by the infinitely many sides of the circles. The sides are no longer strictly fit to measure the lines they touch: they are *lati non quanti*.

There is a delicate question here in interpreting Galileo's remarks about the *lati* and *vacui non quanti*. What is quite clear is that the sides and voids themselves are *non quanti* in the sense that each individual sidelet or void space has no measure; each is an indivisible point that cannot mark a unit of measure of a line or a body. So being *non quanti* is an intrinsic characterization of an indivisible point, whether filled or unfilled. Less clear is whether the characterization of the sides and voids as *non quanti* is also supposed to apply to their being *infinitely many*. Is an infinite multitude, simply by virtue of its infinitude, 'not quantified'? When Galileo says the sides and voids *non son quanti, ma bene infiniti* ("are not quantified, but infinitely many"), this could be taken to contrast being *quanti* with being infinitely many, so that being infinite in multitude is itself a further case of being non-quantified. Or it may instead be taken to clarify the fact that although the sides and voids are *non quanti*—each of them not itself a measurable unit—there are nonetheless infinitely many of them, without thereby implying that *non quanti* applies also the idea of infinite multitude. Which is it?

---

[16] It should be noted that Galileo's analysis is mistaken—at most one of the two circles *rolls* along the tangent, the other merely revolves continuously along it with the illusion of rolling—but our interest concerns the elements of his analysis, not the quality of his solution. For detailed discussion see Drabkin (1950), Costabel (1964), and Knobloch (1999, 2011). See also Mancosu (1996, pp. 121–122).

[17] Knobloch (1999, p. 92).

Unsurprisingly, there are two concepts of quantity to be considered in asking after the meaning of *non quanti* for Galileo, the *metrical* and the *cardinal*. Calling indivisibles *non quanti* is an intrinsic description of points, denying them metrical properties. If *non quanti* is also supposed to characterize infinite multitudes as such, it is then a description denying infinite multitudes cardinal properties. Certainly it is a description denying such multitudes a definite cardinality or number, which would already have been a commonplace view of the time. But perhaps the phrase *non quanti* is further laying the groundwork for denying infinite multitudes very general cardinal properties of comparability: qualifying as 'greater', 'less' or 'equal' in the sense of being more, fewer or equally many. It is this cardinal sense of being *non quanti* that is crucial to asking whether Galileo's reply to the paradox of the natural numbers requires him to deny Euclid's Axiom and the Bijection Principle.

I suspect that Galileo is fairly consistently thinking of *non quanti* through a metrical lens rather than a cardinal one in the passage just reviewed above. After all, he quite explicitly says the line passed over by the smaller circle contains 'just as many' voids (*altrettanti vacui*) as the infinitely many sides (*infiniti lati*) of the smaller circle, even while going on to say that the voids and sides are *non quanti*. A cardinal conception of *non quanti* that disallows comparison of infinite multitudes should have ruled that out. Moreover, the basis for the judgment of equality in that very example is paradigmatically cardinal. In the case of finite polygons, the equal number of sides and voids is established by the one-one correspondence of the sides of the revolving polygons to the parts of the line successively touched or skipped over. In the infinite case of the circles, there will likewise be a succession of touches and skips to establish a one-one correspondence. It is this one-one correspondence which underwrites Galileo's claim that there are just as many voids in the line as sides on the circle, understood as an infinite polygon. So the fact that the sides and voids are *non quanti* does not yet preclude the claim of cardinal comparability. Perhaps when Galileo later denies the possibility of comparison among infinite multitudes, in examining the paradox of the natural numbers, it is not because he thinks infinite multitudes automatically qualify as *non quanti*.

Cardinal notions are at work in his thought as he discusses *non quanti*, and this is evident in the lines that follow immediately on those of the prior passage. Galileo elaborates the idea of composing a finite quantity from an infinity of *non quanti* in order to show how the doctrine of indivisible voids can be deployed to make sense of the possibility of the expansion (and presumably contraction) of lines or solids into spaces of different sizes. Salviati continues with his explanation:

Here I want you to note how, if a line is resolved and divided into parts that are quantified [*in parti quante*], and consequently numbered [*numerate*], we cannot then arrange these into a greater extension than that which they occupied when they were continuous and joined, without the interposition of just as many [*altrettanti*] void < finite > spaces. But imagining the line resolved into unquantifiable parts [*parti non quante*]—that is, into infinitely many indivisibles—we can conceive it immensely [*immenso*] expanded without the interposition of any quantified void spaces, though not without infinitely many indivisible voids.

What is thus said of simple lines is to be understood also of surfaces and solid bodies, considering those as composed of infinitely many unquantifiable atoms [*infiniti atomi non quanti*]; for when we wish to divide them into quantifiable parts [*parti quante*], doubtless we cannot arrange those in a larger space than that originally occupied by the solid unless quantified voids [*quanti vacui*] are interposed—void, I mean, at least of the material of the solid. But if we take the highest and ultimate resolution < of surfaces and solid bodies > into the prime components unquantifiable and infinitely many [*componenti non quanti ed infiniti*], then we can conceive such components as being expanded into immense space [*in spazio immenso*] without the interposition of any quantified void spaces, but only of infinitely many unquantifiable voids [*vacui infiniti non quanti*]. In this way there would be no contradiction in expanding, for instance, a little globe of gold into a very great space without introducing quantifiable void spaces [*spazii vacui quanti*]—provided, however, that gold is assumed to be composed of infinitely many indivisibles. (EN 8: 71–72/D 33–34)

There is much to say about this passage, but we shall focus on just a few points. With the conception of lines and solids as composed of infinitely many *non quanti* indivisibles in mind—a familiar precursor to contemporary point-set analysis of the continuum—Galileo is observing, correctly, how the metrical properties of collections are not directly determined by those of their elements if the elements are allowed to be both infinitely many and to have, individually, no positive measure. The same infinite collection of *non quanti* points might constitute a line of any finite length, or a globe of any size, depending on how the points are arranged. Or more carefully: any assignment of measure might be consistent with a collection of infinitely many *non quanti* points; there is no contradiction in assigning different sizes to such collections.[18] Galileo's appeal to the presence of *non quanti* voids in the lines or solids to explain the differences in measure for the different arrangements seems questionable. It's not clear why *non quanti* voids should expand things any more than *non quanti* atoms would on their own; the appeal to voids seems to serve as a placeholder for whatever it is that makes the difference in the 'arrangement' of the *non quanti* atoms to yield different measures.[19]

The emphasis in most of the passage is on a metrical concept of *non quanti*: the individual points have no measure, and this allows a consistent assignment of any measure at all to lines or bodies composed of them. Cardinality is something of a background condition: there must be infinitely many such *non quanti* points if they are to constitute lines or solids of finite measure in the first place. A finite number of such points cannot suffice. Galileo expressly argues for this claim in response to a different objection to the idea of composing lines from indivisibles. If a line could consist of only finitely many points, it could consist of an odd number of them; but in that case what we might call 'the bisection principle', that a continuous line can always be divided into two equal parts, would require that the middle indivisible be cut, contrary to hypothesis.[20] Galileo replies:

---

[18] This runs parallel to the classical contemporary point-set analysis, which allows unions of infinitely many zero-dimensional points (or singletons) to have any positive measure, though with the proviso, on the contemporary account, that the cardinality of the union be uncountable; countably infinite unions of points would still have measure zero. See Skyrms (1983).

[19] Perhaps an expansion by mere rearrangement of *non quanti* atoms would seem to violate conservation principles, whereas the interposition of *non quanti* voids would not, if void is not a conserved quantity, so to speak.

[20] For provocative discussion of the bisection principle and its possible denial, see Benardete (1964, pp. 240 ff).

> In this, and other objections of this kind, satisfaction is given to its partisans by telling them that not only two indivisibles, but ten, or a hundred, or a thousand do not compose a divisible and quantifiable magnitude [*grandezza divisible e quanti*]; yet infinitely many of them may do so. (EN 8:77/D 39)

Nearly all the uses of the idea of the distinction between *quanti* and *non quanti* in those passages is devoted to the metrical concept. Cardinal properties are involved only in a rather indirect way, when the *non quanti* components are allowed to be infinitely many in order to free up the metrical properties of the finite quantities composed of them. There is no indication of Galileo holding that infinite multitudes may not be judged equal in cardinal terms because they are *non quanti*. The indivisible parts, components, atoms and voids are *non quanti*; the status of infinite multitudes as such remains out of the spotlight.

There is a single phrase at the start of the long passage above, from EN 8:71, that would seem to imply that infinite multitudes cannot be *quanti*, when Galileo describes a line as "divided into parts that are quantified and consequently numbered [*consequenza numerate*]." Taken at face value, this says that being *quanti* entails being numbered, which could well mean having a finite cardinality, rather than, say, just being 'reckoned' into measurable units. If being *quanti* directly implies a finite cardinality in this way, then infinite multitudes will trivially be *non quanti*, and counting an infinite multitude as a 'quantity' would be a 'contradiction in terms', as Knobloch puts it. It is unclear how much weight to assign to this line and to this potential reading of it, but interpreters following Knobloch's lead will want to fasten onto it as evidence that infinite multitudes, just in virtue of cardinality, are automatically *non quanti* for Galileo.

The distinction between *quanti* and *non quanti* comes back to the fore most clearly a little later when Galileo returns to the puzzle of the concentric circles and his solution to it based on the analysis of rotating polygons. Again we can sidestep the details of the analysis and focus on what Galileo says about the limit case of circles:

> If we were to apply similar reasoning to the case of circles, we should have to say that where the sides of any polygon are contained within some number, the sides of any circle are infinitely many: the former are quantified [*quanti*] and divisible, the latter unquantifiable [*non quanti*] and indivisible. (EN 8: 95/D 56)

Although it is clear that the sides of the polygon are finite, *quanti* and divisible, whereas those of the circle are infinite, *non quanti* and indivisible, it is not obvious that *quanti* and *non quanti* refer to more than just the metrical properties of the sides, as one of three distinct categories of properties, roughly: cardinality, measure, and divisibility. And that is, in fact, how I am inclined to read the passage.

For the final reason that leads me to interpret *non quanti* in Galileo as only an intrinsic metrical characterization of indivisibles, and not applying to infinite multitudes just in virtue of their being cardinally infinite, consider a key element of the demonstration of Theorem 1, Proposition 1, in *Two New Sciences*. This proposition is the mean-value theorem for free fall, i.e., the law of falling bodies which says that the time required for an object traveling with uniformly accelerated motion across a given distance is the same as that which would be required for an object traveling

Fig. 2 Diagram for proof of
the mean velocity theorem for
falling bodies. (EN 8:208)

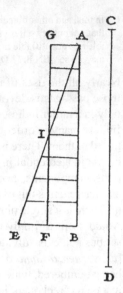

with uniform motion of half the maximum and final degree of speed of the first.
Galileo's proof employs a version of the method of indivisibles, taking aggregates
of 'all parallels'—thus *infinite* aggregates of parallels—cutting across the triangles
and quadrilaterals that contain them. (See Fig. 2.)

Points on the line AB are taken to represent instants of time, and the parallels
drawn from those points across to AIE or GF in the figures in Galileo's diagram represent degrees of speed, either increasing or 'equable.' The parallels do the work of
establishing a one-one correspondence between the instants of time and the degrees
of increasing speed, as well as a one-one correspondence between the instants and
the degrees of equable speed. On the strength of those correspondences, Galileo
is able to conclude that there are "just as many degrees of speed not increased but
equable", and "there are just as many momenta of speed consumed in the accelerated motion as in the equable motion" (EN 8:208–209/D 165–166). 'Just as many'
here is the Latin *totidem*. It is, again, a paradigmatically cardinal treatment of comparison of the aggregates of parallels, despite their being infinitely many. Knobloch
is quite right to observe that Galileo does not treat the parallels as having a *sum*;
Galileo's 'aggregate' does not indicate that the indivisible points or parallels can be
added together.[21] Despite this, the aggregates of points and parallels have cardinal
properties: one-one correspondence implies 'just as many,' despite the fact that the
points and parallels are infinite in multitude.

In my view, Galileo's use of one-one correspondences between infinite multitudes to justify cardinal claims of equality—the non-numerical but precise claim
that there are just as many Xs as Ys—is the strongest evidence that Galileo does
not regard infinite multitudes *per se* as falling under the rubric of *non quanti*, which

---

[21] Knobloch (1999, p. 93; 2011).

seems to be primarily a metrical concept rather than a cardinal one. There is a risk in this argument, of course, since Galileo's claims of cardinal equality between infinite multitudes also appear to run directly contrary to the very philosophical pronouncement that is driving our present inquiry, namely, that "one infinity cannot be said to be greater or less than or equal to another" (NE 8: 78/D40).[22] It may thus 'prove too much' to appeal to Galileo's mathematical practice, when evaluating his considered philosophical position. The practice in some places appears to be patently at odds with the philosophy, even apart from the question of *non quanti*; perhaps it is incautious to draw a philosophical conclusion from it.

In fact, however, the cases are not truly parallel. The clash between the proof of *Thm.* 1, *Prop.* 1, described on the Third Day and Salviati's denial of comparability among infinites during the First Day's discussion of paradoxes at least takes place across relatively distant points within *Two New Sciences*. And of course the 'mathematical demonstrations' are attributed to 'the Author' of the Latin treatise *On Local Motion*, i.e. Galileo, and not necessarily endorsed in every respect by the more philosophically drawn Salviati. As we saw above, however, Galileo's (Salviati's) assertions of cardinal equality among infinite multitudes of indivisibles—between the sides and the voids—occur even in the same breath as his careful efforts to distinguish *quanti* from *non quanti*. If infinite multitudes are supposed to be cardinally incomparable on grounds of being *non quanti*, Galileo's analyses of the rolling circles and polygons in those texts are then grossly mistaken on their own terms. This strikes me as a needlessly damaging interpretation. A more natural and less destructive reading is to take *non quanti* as a metrical concept that is not meant to cover just any case of infinite multitude, and further to take the denial by Salviati, later in the First Day, of comparability among infinites to be based on a somewhat different mix of considerations, which we shall consider shortly below.

To sum up, Galileo's distinction between *quanti* and *non quanti* appears to be metrical rather than cardinal in character. While it marks out a crucial difference between the intrinsic properties of divisible parts and those of indivisible ones—only the former are *quanti* and suitable for 'measure'—it does not by itself rule out judgments of cardinal equality among infinite multitudes.

Before pressing ahead with this result, it is worth noting that we should not be too quick to pull apart the metrical and cardinal concepts of quantity in Galileo, as if they were wholly severable in his thought.[23] The idea of cardinality as something determined by one-one maps—by *functions* on *sets*—would not be properly

---

[22] Another possibility is that Galileo's denial that 'greater', 'less' and 'equal' apply in the infinite is carefully consistent with his judgments that there as just as many Xs as Ys in some cases of infinite multitudes: perhaps the Xs and the Ys can be *just as many* without falling under the term 'equal'. If so, however, it would seem to be only a matter of a word, as no richer notion of cardinal equality seems available for which 'just as many' is not sufficient. A more substantial possibility here would be that for Galileo, 'greater', 'less' and 'equal' are essentially metrical notions, and their cardinal counterparts 'more', 'fewer' and 'equally many' cannot be applied on the basis of one-one maps without corresponding geometrical judgments in place as well. I am more sympathetic to this idea but cannot pursue it here; a few related points are discussed below.

[23] I am indebted here to Katherine Dunlop.

distilled until well into the eighteenth century or even later. Galileo's handling of one-one relations between elements of multitudes, especially infinite multitudes, is typically mediated by geometrical relations between the mathematical objects that contain those elements. In the case of Aristotle's wheel, for instance, the one-one correspondences between sides of the rotating polygons (including the circles) and parts and voids in the lines are established under the aegis of geometrical relations between sides of the polygons and the parts of the lines they touch. For instance, the judgment that the line passed over by the smaller circle contains as many voids as there are sides on the circle is based on the following consideration. For every length of the line BF passed over by a side of the larger circle, an equal segment of the equal line CE must be passed over by a side of the smaller circle. (See, again, Fig. 1.) Since each side of the smaller circle is shorter than the segment of the line CE that it passes over, the side touches only some part of that segment; thus there must a be void interval remaining in that segment which the circle 'skips over' in its passage. (Why must each side of the smaller circle be shorter than those of the larger circle? Because their ratio is supposed to be preserved when we shift from the case of finite polygons to the case of infinilateral circles.) The one-one correspondences among the elements of the figures—sides, parts, voids—are thus fixed within a pattern of systematic geometrical relations between the figures themselves.

Nowadays measurements of cardinality require that the two objects compared are *sets*, and this remains a natural counterpart (and perhaps remnant) of the earlier thought that the two measured objects are wholes. The measurements themselves are effected by functions, which need not require any rule or procedure by which to relate the elements of the two sets. In the early modern context, however, the idea of a completely *arbitrary* relation between the elements—a purely 'combinatorial' concept of function—would have been rather alien. Thus measurement of cardinality still needed to be supplemented by other considerations, ones in which metrical concepts often had important roles to play. When we find, as we shall below, Galileo limiting judgments of cardinal equality among infinite multitudes to cases with specific metrical constraints—namely, that the objects be *bounded* quantities—there should be no surprise. This arises organically from the way in which geometrical concepts remain coeval with more purely arithmetical ones in his thought.

1.5   Euclid's Axiom Revisited: Infinity, Magnitude and Infinite Number in Galileo

If our conclusions in the last section are right, then when Galileo denies comparability among infinite classes of numbers, no simple appeal to the distinction between *quanti* and *non quanti* will offer a route of escape from overturning the Bijection Principle. Infinite multitudes are not automatically *non quanti*, at least with respect to cardinal comparisons, and therefore they are not out of play as potential counterexamples to the Bijection Principle. Likewise, Knobloch's original proposal that Euclid's Axiom is not invalidated on Galileo's account because infinites are *non quanti* seems not to be fully sustained either. If the various infinite number classes are *non quanti* and thereby outside the scope of Euclid's Axiom, that remains to be shown; it does not follow merely from their being infinite in multitude.

Yet Knobloch's defense of Galileo in the case of Euclid's Axiom strikes very close to the truth. The most natural concept of size for the idea that the whole is greater than the part is a metrical one rather than a cardinal one. So an extension of Galileo's distinction between *quanti* and *non quanti* parts of lines or solids to multitudes of those parts could comfortably rule that only multitudes with a finite total measure can count as *quanti* and thus qualify for comparison. For whereas the concept of infinite multitude might admit of precise mathematical handling in terms of maps, classes, etc., the concept of infinite measure—infinite *magnitude*, as we might say—is less amenable to mathematical analysis and would arouse more skepticism. And Galileo himself seems to confine mathematical analysis to objects that are in some way limited in magnitude. As Knobloch notes, Galileo's term for this is *terminata* or 'bounded.'[24] When considering a line or a circle or a solid as composed of infinitely many indivisibles, it qualifies as a quantity fit for mathematical treatment only if it is itself a bounded magnitude. Galileo remarks on the "infinite difference and even repugnance and contrariety of nature in a bounded quantity in passing over to the infinite" (EN 8:83/D 46):

> Consider, then, what a difference there is <in moving from> a finite to an infinite circle. The latter changes its being so completely as to lose its existence and its possibility of being <a circle>. For we understand well that there cannot be an infinite circle, from which it follows as a consequence that still less can a sphere be infinite; nor can any other solid or surface having shape be infinite. What shall we say about this metamorphosis in passing from finite to infinite? (EN 8:85/D 47)

In fact Galileo does not say exactly what his answer to this question is, beyond the idea that the nature of the objects in question changes or is lost entirely in passing to the infinite, and, in his earlier words of admonition, "These are among the marvels that surpass the bounds of our imagination and that must warn us how gravely one errs in trying to reason about infinites by using the same attributes that we apply to finites; for the natures of these have no necessary relation between them" (EN 8:83/D 46). But the warning is clearly about passing from the intelligible to the unintelligible, and as Knobloch keenly observes, Galileo's language and argumentation clearly evoke that of Nicholas of Cusa, who originated the distinction between (in the Latin) *quanta* and *non quanta* and imbued the whole topic with almost mystical significance.[25] If Galileo does not quite say that the infinite circle and sphere are *non quanti*, the allusions to Nicholas may in effect do this for him.

With that in mind, a second look at the lesson Galileo draws from the paradox of the natural numbers readily finds the same concern to limit comparisons to bounded quantities: "And in the final conclusion, the attributes of equal, greater and less have no place in infinite, but only in bounded quantity [*quantità terminate*]" (EN 8:79/D

---

[24] Knobloch (1999, p. 92). See also Knobloch (2011).

[25] See Nicholas ([*1440*] 1985), *De Docta Ignorantia*, Book 1, Chaps. 11–23; see especially Chap. 14 for use of the distinction between *quanta* and *non quanta*.

41). This is revealing, for it shows that Galileo sees the crux of the problem to be the unboundedness of the multitude of natural numbers. His contrast between 'infinite' and 'bounded' strongly suggests that he is thinking of magnitude rather than multitude. 'Bounded' of course has cardinal as well as metrical senses, but it would be redundant here to point out that the infinite multitude of natural numbers is cardinally unbounded. By contrast, pointing out that the natural numbers taken together are infinite and unbounded in magnitude distinguishes them in a special way, for an infinite multitude is not always unbounded in magnitude. Even on Galileo's view, infinitely many points may compose a bounded quantity such as a finite globe. Because the individual points themselves are *non quanti*, taking infinitely many of them in aggregate need not total up to an unbounded quantity; they can consistently be taken to compose a bounded quantity of finite magnitude such as a little globe of gold. The same could not be true of *quanti* parts, unless they happen to form a convergent infinite series (e.g., $1/2 + 1/4 + 1/8 + 1/16$, &c., whose sum is equal to 1). Apart from this exceptional case (which was much disputed at the time and is not mentioned by Galileo in connection with the topic of *quanti* and *non quanti*), any infinite collection of finite quantities will surpass any finite size and thus be unbounded.

When Galileo holds the multitude of natural numbers to be infinite and unbounded, it seems he is regarding each natural number as if it were a 'quantifiable' part of all the naturals. This is not so strange an idea, and it shows that the concept of number at work includes metrical as well as cardinal elements. Taking the natural numbers to be multiples of the number one as the basic unit of 'arithmetical measure', each natural itself is a finite quantity. If all the naturals are taken together there is no way for the total to be bounded, i.e. less than or equal to some finite magnitude, i.e. the magnitude of some finite number. Their aggregate would instead appear to be an infinite magnitude, as if a colossal infinite number made up of all the finite numbers or infinitely many copies of the unit number one.

If this is how the infinity of natural numbers is understood by Galileo, then there is a case to be made that it too qualifies as 'non-quantifiable', like the infinite globe or infinite circle, not in virtue of its smallness but in virtue of its greatness. But it is metrical greatness—greatness of magnitude—rather than cardinal greatness that yields the verdict of *non quanti*. For there is no cardinal difference between the infinity of natural numbers and the infinity of indivisibles in a little globe of gold, or the infinity of sides of a circle. There is textual evidence that Galileo thinks of the idea that the natural numbers are susceptible to comparative measurement as representing the natural numbers as components in a single colossal number with infinitely many finite parts. And it comes directly in his discussion of the paradox of the natural numbers. Having established the one-one correspondences among squares, roots and naturals, Galileo notes:

> That being the case, it must be said that square numbers are as numerous as all numbers, because they are as many as their roots, and all numbers are roots. Yet at the outset we said that all the numbers were many more than the squares, the greater part [*maggior parte*] being non-squares. Indeed, the multitude of squares diminishes in ever-greater ratio as one moves on to greater numbers, for up to one hundred there are ten squares, which is to say one-tenth part [*parte*] are squares; in ten thousand, only one-hundredth part [*parte*] are squares; in one million, only one-thousandth. (EN 8:78–79/D 40–41)

The farther the series of numbers is finitely extended, the smaller the share of squares becomes, so it seems especially peculiar how in the infinite case a transition occurs to make the squares and naturals equal. Now notice exactly how Galileo phrases his point in the line that follows:

> Yet in the infinite number [*numero infinito*], if one can conceive that, it must be said that there are as many squares as all numbers together. (Ibid.)

The comparison between the squares and the naturals that shows them to be equal takes place 'in the infinite number', of which, it seems, the squares form a share or part. So it appears that comparison of infinite multitudes of numbers, at least, presupposes a single number after all—not in order to stand as the absolute cardinality of the multitudes being compared, but to be a common object of which each multitude is some proportional part and can be compared in ratio to the other. The common object in this case is a number because it is composed of numbers, the infinite aggregate of all numbers. Perhaps if comparisons were being made among infinite multitudes of other types, the need for a common object would not automatically entail the existence of an infinite number but of something else, say, the aggregate of all stars in the heavens, or, to use a case straight from Galileo's texts, the aggregate of all parallels in a given figure or points in a line or sides on a circle. Galileo is not explicit about the details. But it is clear enough in the case of the natural numbers that comparisons among its infinite subclasses are conceived by Galileo to involve an infinite number as the common object containing the compared classes as parts.

The same strand of thought is carried through when Galileo returns to the paradox later in the dialogue. It appears this time when Salviati introduces the revisionary suggestion that infinity be regarded, if anything, not as a number standing at the far end of the natural numbers but instead as the number one:

> In our discussion a little while ago, we concluded that in the infinite number [numero infinito], there must be as many squares or cubes as all the numbers because both <squares and cubes> are as many as [tante...quanti] their roots, and all numbers are roots. Next we saw that the larger the numbers taken, the scarcer became the squares to be found among them, and still rarer, the cubes. Hence it is manifest that to the extent that we go to greater numbers, by that much and more do we depart from the infinite number. From this it follows that turning back (since our direction took us always farther from our desired goal), if any number may be called infinite, it is unity. And truly, in unity are those conditions and necessary requisites of the infinite number. I refer to those <conditions> of containing in itself as many squares as cubes, and as many as all the numbers <contained>. (EN 8: 82–83/D 45)

Leave aside the new suggestion that the infinite number is unity or one. What matters for us is Galileo's consistent appeal to the idea of an infinite number in his handling of comparisons between the squares, cubes, roots and naturals. For these classes of numbers to be compared to one another, it seems as if they must be able to stand in ratio to one another and to form parts or shares of some single common whole, the infinite number formed out of all the natural numbers taken together (at least prior to Salviati's revisionary identification of the infinite number with unity). Such a number would indeed be infinite in magnitude, unbounded, and comfortably regarded as 'non-quantifiable' on account of its greatness.

It is instructive to note that in the instances in which Galileo relies on comparisons of equality among infinite multitudes in his avowed mathematical arguments, those multitudes are confined to bounded spaces of finite magnitude. His proof of Thm. 1, Prop. 1 in *On Local Motion*, for example, trades on the cardinal equality of degrees of speed and instants in time; but these 'infinites' are represented as aggregates of points or parallels within bounded geometrical figures. Unbounded magnitude is in effect ruled out from the start, and with that safe harbor established it seems Galileo is prepared to make use of cardinal comparison, or at least cardinal equality, among infinites. 'Safe harbor' perhaps suggests too much, for we have not seen whether it is a *sufficient* condition for infinite collections to be cardinally comparable that they be bounded in magnitude. The present suggestion is only that Galileo seems to treat it as a *necessary* condition, one which the aggregate of parallels in a bounded figure satisfies but the collection of all natural numbers does not. What else, precisely, is required for a sufficient condition for cardinal comparability is not immediately clear; though as suggested before, perhaps some geometrical relation or finitary rule or procedure would be expected in order to establish the one-one mapping among the elements.

The result of all this would appear to be that, given Galileo's reliance on the idea of an infinite number as a common whole in framing comparisons involving all the natural numbers taken together, his distinction between *quanti* and *non quanti* can be extended to count the infinity of natural numbers as 'non-quantified'—not because they are infinite in multitude but because the infinite number they compose would be unbounded in magnitude and thus unfit for comparison. This then also preempts the natural numbers, and any similar infinite class of numbers, from constituting a potential counterexample to Euclid's Axiom. Such infinites, by virtue of their magnitude, are *non quanti* and unfit for comparison, and thus do not fall within the scope of Euclid's Axiom in the first place. So with respect to the specific paradox of the natural numbers, Knobloch's diagnosis appears to be correct.

Yet not all infinite multitudes will likewise amount to infinite magnitudes when taken all together, and those which do not, such as infinite multitudes of indivisibles in bounded lines or solids, may still be open for comparison in cardinal terms by means of one-one maps. There are as many sides in the circle as indivisible parts in the line it traverses. The difficulty now reappears just as it emerged in Simplicio's original objection. If finite and bounded quantities consist of infinitely many indivisible points—their *non quanti* parts—examples can easily be found in which the whole is equal to the part, by use of one-one correspondences and the Bijection Principle. Parallels can be dropped from all the points on the diagonal of a square to all those on one of its sides, and yet by rotating that side back up to the diagonal, the points on the side can be put into one-one correspondence with those of only a part of the diagonal. Thus all the points on the diagonal can be put into one-one correspondence with the subclass of those points composing the part of it congruent with the rotated side.[26] Again Euclid's Axiom and the Bijection Principle conflict: the diagonal is both greater than and equal to its part.

---

[26] This so-called Diagonal Paradox was well-known by the seventeenth century; see Leibniz's use of it at A VI, 3, 199. As noted by Lison (2006/2007, p. 199fn3), the example goes back at least to

If Galileo's denial of comparability is limited to cases in which infinite multitudes constitute aggregates of infinite magnitude, there is no escape from contradiction in the finite bounded case. But if his denial is taken to apply across the board even to those infinite multitudes that compose only bounded magnitudes, then it seems either Euclid's Axiom or the Bijection Principle (or both) must be a casualty of his analysis after all.

Similarly, a limited denial of comparability of infinites would allow Galileo's use of the Bijection Principle in his mathematical practice to remain in harmony with his philosophical pronouncements, since he appears to confine that use to infinite multitudes housed within bounded quantities. But an across-the-board denial of comparability among infinites will yield a clash with his mathematical practice of applying the Bijection Principle in reasoning about infinite classes of indivisibles in the bounded cases.

The lesson of our inquiry is that the distinction between *quanti* and *non quanti* will not allow Galileo a fully reconciled position that avoids contravening at least some major mathematical principle of comparison. If all infinite multitudes are *non quanti*, then Galileo's ready use of cardinal comparison among infinite multitudes in bounded cases is, as we have seen, grossly at odds with his philosophy on this point, even in the very texts in which he articulates his notion of *non quanti*. On the other hand, if infinite multitudes are *non quanti* only when they constitute unbounded magnitudes—as I have been suggesting—then paradox reemerges in cases in which infinite multitudes compose only finite and bounded magnitudes, and Galileo has no choice but to abandon either Euclid's Axiom or the Bijection Principle.

Unless, that is, there is altogether another way for him to solve paradox.

1.6 Leibniz's Alternative

In notes written in 1672 on *Two New Sciences*, Leibniz reviews Galileo's paradox of the natural numbers:

> He [Galileo] thinks that one infinity is... not greater than another infinity... And the demonstration is worth noting: Among the numbers there are infinite roots, infinite squares, infinite cubes. Moreover, there are as many roots as numbers. And there are as many squares as roots. Therefore there are as many squares as numbers, that is to say, there are as many square numbers as there are numbers in general [*in universum*]. Which is impossible. (A VI, 3, 158)

Leibniz sees that the equality of the squares with the natural numbers will violate Euclid's Axiom and that one option is simply to reject the axiom in the case of the

---

Ockham. Note also that similar examples can be constructed even with multitudes containing only *quanti* parts, provided their total measure remains finitely bounded. Considering a line segment as composed of a sequence of geometrically decreasing non-overlapping subsegments, one can easily construct a one-one correspondence between the subsegments of the side of a square and the subsegments of the diagonal, even though the side can be shown by rotation to be equal to a part of the diagonal.

infinite, which he takes to be Galileo's solution. Yet, as we observed once before, he regards this as unacceptable, an admission that the axiom itself has counterexamples, and he appears to see another way out of the paradox:

> Hence it follows either that in the infinite the whole is not greater than the part, which is the opinion of Galileo and Gregory of St. Vincent, and which I cannot accept; or that infinity itself is nothing, i.e. that it is not one [*Unum*] and not a whole [*totum*]. (Ibid.)

What is Leibniz's alternative to (what he takes to be) Galileo's answer? First let us understand the elements of his reply.[27] When Leibniz denies that infinity is one or a whole he is not saying that there is no such thing as infinity, but rather he is denying that an infinity of things forms a unity or single whole. It is not the number one (contrary to Salviati's revisionary suggestion), nor is it to be understood as a single set with infinitely many elements. Leibniz's position here is subtle. There are actually infinitely many natural numbers, on his view, but they do not form a totality. There is only a plurality or multitude of them, but no one thing, no single object, to which they all belong as constituents. There is no set of all natural numbers, so to speak, but only a proper class. When Leibniz says "infinity itself is nothing" he is not denying that there are infinitely many numbers; he means there is nothing over and above the natural numbers themselves. To call them 'infinite' is not to posit a special entity—a super-number or super-whole embracing all the naturals and larger than every finite number. It is, rather, to describe the multitude of natural numbers in a particular way. Exactly what he means by 'infinite' we shall discuss shortly below.

It is fairly straightforward to see how this can solve the paradox. By denying that an infinite multitude can form a whole, Euclid's Axiom is taken out of play, and along with it the consequence that there must be more naturals than squares. The infinite multitude of natural numbers is not a whole of which the square numbers form a part, so the infinity of natural numbers does not have to be said to be greater than the infinity of the squares. This removes the contradiction that the naturals are both greater than and equal to the squares. Note that it also does so without contravening Euclid's Axiom. If there are no infinite wholes, then there is no whole that can fail to be greater than the part, and so no counterexample to the axiom.

It is less clear whether infinite multitudes can or cannot be compared to one another at all on Leibniz's view. Unlike the earlier strategy of looking to classify infinites as *non quanti* and thereby preclude them from comparison altogether, denying that infinites form wholes only takes them out of the specific jurisdiction of Euclid's Axiom. It remains open, at least, to compare them using one-one maps and the Bijection Principle. What comparisons, if any, does Leibniz allow under this principle? We will not resolve the issue completely in the present essay, but a few points may be instructive nonetheless.

I think the matter turns on what is involved in the idea of there being 'as many' elements in one class as in another. As before, one thought is that in order for there to be as many Fs as Gs, the must be a *number* that records how many Fs there are,

---

[27] For related discussion, see Levey (1998) and Arthur's introduction to Leibniz (2001).

and likewise a number to record the cardinality of the Gs. If those two numbers are equal, then there are as many Fs as Gs, and not otherwise. For reasons we shall not pursue, however, Leibniz holds that there is no such thing as an infinite number, no number that could say of an infinite class just how many elements there are in it.[28] To call a class infinite is not to assign it a number, on his view. If infinite classes are without number, and if there are as many Fs as Gs only if they are the same in number, then it cannot be said that there are as many Fs as Gs when they are infinite. There may well be one-one correspondences between the Fs and the Gs, but this will not automatically yield the conclusion that they are equal in any sense beyond the idea that both classes are infinite.

Leibniz does not, to my knowledge, explicitly say that equality of class size—in the sense of 'as many Fs as Gs'—requires equality of number. But it is clear that he thinks the question of equality of number is relevant to the analysis of Galileo's paradox. For in a subsequent discussion of the paradox, in the dialogue *Pacidius to Philalethes* (1676), he systematically frames the issue not in terms of there being *as many* squares as numbers, as he had in the earlier notes on *Two New Sciences*, but in terms of 'the number of all squares' and 'the number of all numbers' (A VI, 3, 550). The interlocutors consider Galileo's response to the paradox and reject it:

> CHARINUS: Please allow me to hear first from Gallutius what Galileo said.
> GALLUTIUS: He said: the appellations 'greater', 'equal' and 'less' have no place in the infinite.
> CHARINUS: It is difficult to agree with this. For who would deny that the number of square numbers is contained in the number of all numbers, when squares are found among all numbers? But to be contained in something is certainly to be a part of it, and I believe it to be no less true in the infinite than in the finite that the part is less than the whole.
> (A VI, 3, 551/Ar 178–179)

The commitment to Euclid's Axiom is clear, as is the deployment of the language of parts and wholes in describing the relation between the squares and the natural numbers. In fact the view here seems to integrate 'number', 'whole', 'part' and 'containment' quite fully. Not only are the square numbers among all the numbers, but also the *number* of squares is a *part* of the number of all numbers; the latter number is the whole that contains the part. Leibniz's strategy for solving the paradox by denying that the natural numbers forms a whole is thus expressed in subsequent lines by Charinus through the claim, "there is no number of all numbers at all, and that such a notion implies a contradiction" (ibid.). The denial that there is such a number is the denial that the natural numbers form a whole. Leibniz's principal spokesman, Pacidius, then praises Charinus's answer as "very clear, and if I am any judge, true",

---

[28] Leibniz has a few different lines of argument to offer against infinite number, including, notably for us, a deployment of Galileo's paradox that expressly says infinite numbers are "impossible" because "it is impossible that this axiom"—Euclid's Axiom—"fails" (A III, 1, 11). (This passage comes from some remarks by Leibniz on Galileo's paradox, written in 1673.) Here an infinite number is presumably understood as a whole constituted of infinitely many units or the combination of all natural numbers, themselves taken as wholes composed of units. For some related discussion, see Levey (1998).

and adds, "for it is necessary that what has contradictory consequences is by all means impossible" (ibid.)

Again, Galileo does not quite say directly that the natural numbers form a whole; Leibniz is reading this into Galileo's position. But as we saw, Galileo's account of what it would be for the natural numbers to be comparable to the squares or cubes, etc., seems to involve the existence of an infinite number, made up of all the numbers, as the common object of which the squares form a part. So while Galileo does not say that the (cardinal) equality of two infinite multitudes must be equality in number, in the sense that there is some single number that says precisely how many elements there are in each multitude, nonetheless for the case of comparison of infinite number classes his analysis does seem to require the existence of an infinite number in which those classes are contained. It is not evident that Leibniz has discerned all this in Galileo—he does not describe Galileo's account in enough detail to tell—but when he attributes to Galileo the idea that the natural numbers form a whole, it seems he has hit upon the truth.

Leibniz's solution to the paradox does not require him to reject Euclid's Law. And it seems also to leave the Bijection Principle intact, if only because the principle is silent about the distinction between whole and multitudes, and silent also about the concept of number, and so is not called into question by Leibniz's denial of infinite wholes and infinite numbers. It remains to ask whether Leibniz himself will endorse the Bijection Principle or any related form of comparability of infinite classes using one-one maps; we turn to that in the next section. Before doing so, however, it should be noted that Leibniz's solution does not completely disarm Galileo's paradox .

Just as the appeal to the distinction between *quanti* and *non quanti* left a version of the paradox untouched when reconfigured from an unbounded infinite case to a bounded case—for instance, the Diagonal Paradox—so too there remains a problem for Leibniz's strategy. For in order to prevent Euclid's Axiom and the Bijection Principle from coming into conflict by appeal to the distinction between wholes and pluralities, he will have to deny that *any* infinite multitude forms a whole. This goes for bounded and unbounded infinite multitudes alike. In mathematical examples of the unbounded case, this means that the classes of natural numbers, of squares, etc., are not wholes. In a natural-world example, this means that the infinite universe itself does not form a whole—a conclusion Leibniz expressly draws on the basis of his analysis of Galileo's paradox.[29] But both mathematics and the natural world also provide bounded cases of infinite multitudes: the Diagonal Paradox for finite, bounded geometrical figures; and any given finite body in nature, since according to Leibniz every body is actually infinitely divided into parts.

---

[29] Leibniz writes: "*God is not the soul of the world* can be demonstrated; for the world is either finite or infinite. If the world is finite, certainly God, who is infinite, cannot be said to be the soul of the World. If the world is supposed to be infinite, it is not one Being or one body *per se* (just as it has elsewhere been demonstrated that infinite in number and in magnitude is neither one nor a whole, but infinite in perfection is one and a whole). Thus no soul of this sort can be understood. An infinite world, of course, is no more one [Being] and a whole than an infinite number, which Galileo has demonstrated to be neither one nor a whole" (Leibniz 1948, p. 558).

In the case of geometrical figures, it is easy enough for Leibniz to accept the verdict that they are not truly wholes. After all, they are, on his view, only *entia rationis* and not real beings. The same conclusion is harder to accept in the case of bodies, for it will quickly preclude any body from truly being a single whole, calling into question its reality and indeed the reality of the entire corporeal world. Certain interpretations of Leibniz might welcome this result, while others would find it an unhappy fit with his views at least for important parts of his career.[30] Leaving aside the implication for 'idealist' and 'realist' interpretations of Leibniz, however, it should be noted that Leibniz shows no sign of intending his analysis of Galileo's paradox to issue in a denial of the reality of finite corporeal beings. Since Galileo, like Leibniz, regards bodies as infinitely divided into parts, if he were to take Leibniz's path of escape from overturning Euclid's Axiom and the Bijection Principle, he too would face this discomforting consequence for the natural world. Thus although Leibniz's distinction between wholes and multitudes might provide a way to resolve the contradiction in mathematics alone, the paradox continues to hold some hostages in metaphysics.

1.7   Comparability and the Definition of Infinite in Leibniz and Galileo

There are a few last pieces of Leibniz's view that still need to be articulated, concerning comparability, number and infinity. Leibniz rejects both infinite totalities and infinite numbers. Does he also reject the idea that infinite classes can be comparable? The squares and the naturals cannot literally be equal *in number*, and they cannot be equal parts of a single common object, the infinite number. But might they nonetheless be equal in the general sense of being 'just as many', in virtue of there being a one-one correspondence between them? That is, does Leibniz endorse the Bijection Principle even for infinite classes?

In at least one clear statement, first noted by Russell, Leibniz denies the claim of equality for infinite classes despite the existence of a one-one correspondence between them:

> There is an actual infinite in the mode of a distributive whole, not of a collective whole. Thus something can be enunciated concerning all numbers, but not collectively. So it can be said that to every even <number> corresponds its odd <number>, and vice versa; but it cannot be accurately said that the multitudes of odds and evens are equal. (GP II, 315)

The link between speaking of 'all numbers' and taking them as a 'collective whole' is explicit, and it seems that equality of the odds and the evens has to be denied precisely because such infinite classes do not form collective wholes (presumably 'distributive whole' just means multitude or plurality here). Leibniz's solution to Galileo's paradox protects Euclid's Axiom but in turn leads him to reject the Bijection Principle for the infinite case—*if*, that is, this passage reflects his considered

---

[30] For an overview of this dispute in the interpretation of Leibniz, see Levey (2011).

position. It should be noted that the passage occurs on a separate slip of paper enclosed in Leibniz's copy of his letter of 1 September 1706 to Bartholomew Des Bosses, and Leibniz has crossed it out.[31] It is hard, therefore, to know how much weight to give it.

Even if it were Leibniz's view that the odds and evens are not equal, it cannot be the entirety of Leibniz's position to deny comparisons involving infinite classes. For he has to preserve at least one crucial claim of comparability for infinite classes, namely, that there are more elements in an infinite class than in any finite one. This is essential to Leibniz's definition of infinite:

> When it is said that there are infinitely many terms, it is not being said that there is some specific number of them, but that there are more than any specific number. (GM III, 566)

This is no fleeting aspect of Leibniz's philosophy of mathematics but his deeply held definition of 'infinite' with ramifications across his thought, and thus he is quite committed to the coherence of this comparison between infinite and finite classes. Interestingly, Galileo, in the character of Sagredo, denies that infinite quantities can be compared even with finite ones, apparently in an echo of Aristotle's prohibition, in *De Caelo* 274a10 and 274b12, against ratios between finite and infinite; cf. EN 8: 79–80/D 42.[32] (Again, though, this may be limited to comparisons of magnitudes and not to multitudes.)

Taken just at face value, and if we include the crossed-out remark on the slip in the letter to Des Bosses, Leibniz may seem to have landed in almost the inverse of Galileo's position: infinite multitudes cannot be said to be equal, but the infinite can be compared to the finite and judged to be (cardinally) greater. The particular example Leibniz offers in denying equality among infinites is one about which Galileo would agree. For the evens and the odds would each constitute an unbounded infinite magnitude if taken all together and thus fail to be *quanti* for Galileo, thereby preempting a judgment of equality.

Unlike Galileo, whose denial of comparability among infinites turns out to be linked to the concept of magnitude—infinite magnitudes are unbounded and thus *non quanti* and incomparable—Leibniz's denial of comparability (or at least equality) among infinites relies only on the distinction between 'collected' wholes or totalities and multitudes, without any obvious reference to the concept of magnitude. So whereas Galileo can allow infinite multitudes of finite and bounded magnitude to be among the *quanti* and hence comparable, it is not clear that Leibniz can make the parallel allowance that those infinite multitudes of finite and bounded magnitude can count as wholes. Given similar opportunities to treat bounded mathematical magnitudes with infinitely many elements as wholes, he appears to take steps not to do so.[33]

---

[31] See Leibniz (2007), p. 409.

[32] Drake suggests that Galileo neither fully accepted nor fully rejected Aristotle's principle (1974, 42fn26).

[33] See Levey (1998, 1999, 2003). For defense of the view that Leibniz's strictures against infinite wholes only preclude wholes of infinite magnitude, and thus can allow wholes of finite magnitude that include infinitely elements, see Brown (2005).

Either Leibniz has an unstated subtlety at work here in his restrictions on comparability—one that allows the claim of 'more than' between infinite and finite classes but rules out the claim of equality among the evens and odds—or he is inconsistent in his statements across texts. I am inclined to think the denial of cardinal equality in the crossed-out passage is an outlier and not Leibniz's considered view. As noted earlier, however, this matter will not be entirely resolved here. In any case, for now it is enough to note that insofar as Leibniz requires comparability of an infinite class with a finite one for his definition of 'infinite', infinite multitude is not *eo ipso* a barrier to it.

On the definition of 'infinite' there is a further illuminating, and perhaps ironic, connection between Galileo and Leibniz. In *Two New Sciences*, Galileo has the interlocutors consider the question of how many *quanti* parts a bounded continuum such as a finite line segment can be divided into. Finitely many or infinitely many? Salviati's answer is that there are neither finitely many nor infinitely many, but something intermediate between the two:

> *Salviati*: To the question which asks whether the quantified parts in the bounded continuum are finite or infinitely many [*infinite*], I shall reply exactly the opposite of what Simplicio has replied; that is, 'neither finite nor infinite.'
>
> *Simplicio*: I could never have said that, not believing that any middle ground is to be found between the finite and the infinite, as if the dichotomy or distinction that makes a thing finite or else infinite were somehow wanting and defective.
>
> *Salv.*: It seems to me to be so. Speaking of discrete quantity it appears to me that there is a third, or middle, term; it is that of answering to every [*ogni*] designated number. Thus in the present case, if asked whether the quantified parts in the continuum are finite or infinitely many [*infinite*], the most suitable reply is 'neither finite nor infinitely many, but so many as [*ma tante che*] to correspond to every specified number.' To do that, it is necessary that these be not included within a limited number, because then they would not answer to a greater<number>; yet it is not necessary that they be infinitely many, since no specified number is infinite [*infinito*]. And thus at the choice of the questioner we may cut a given line into a hundred quantified parts, into a thousand, and into a hundred thousand, according to whatever number he likes, but not into infinitely many<quantified parts>.
> (EN 8: 81/D 43–44)

Galileo's 'middle term' between finite and infinite is almost exactly Leibniz's official definition of infinite, differing only in whether the parts in the multitude are so many as to *correspond to* any specified number or to be *more than* any specified number. This quickly comes to the same thing, since, as Galileo points out, corresponding to any specified number requires not being included in any limited number; so for any given number n, the parts in the multitude must exceed n anyway. In his 1672 notes on this passage, Leibniz just describes this as saying that the parts in the continuum are 'indefinite.' No doubt this captures some of Galileo's intention in placing emphasis on the 'choice of the questioner.' But the relation between the variables here—between the number chosen and the comparison of the multitude with that number—is no mere matter of indefiniteness, and it later becomes the key to Leibniz's considered definition.

As is now recognized among commentators, Leibniz advances a *syncategore-matic* analysis of the term 'infinite' rather than a *categorematic* one.[34] To say that a class is infinite is not to assign it a single infinite number but to describe it in terms of logical relations among variables referring to finite numbers: the Fs are infinite iff for any n, there are more than n Fs. By contrast, a categorematic analysis of 'infinite' would make reference to an infinite number: the Fs are infinite iff there is a number k of Fs such that k is greater than any finite number n. (The order of quantifiers is crucial of course.) This is not wholly an innovation by Leibniz. The distinction between categorematic and syncategorematic terms goes back at least to Priscian, and the use of this distinction in the analysis of the term 'infinite' predates Galileo and Leibniz by a few centuries.[35] What is notable here for us is that Galileo clearly reveals his own interpretation of 'infinite' to be a categorematic one. When he writes of the multitude of *quanti* parts in a bounded continuum that they are so many as to correspond to any specified number 'yet it is not necessary that they be infinite, since no specified number is infinite,' he is expressly holding that for the multitude to be infinite it must correspond to an infinite number—the unmistakable signature of the categorematic account. What Galileo takes to define an intermedi-ate status between infinite and finite is exactly what Leibniz later appropriates for his own syncategorematic analysis of 'infinite'.

A last point is important to bring out in Leibniz's account. Taking care with his statement of what is meant by saying there are infinitely many terms, very precise definitions of 'finite' and 'infinite' can be formulated. A class X is infinite if and only if for any number n, there are more than n elements of X. And by negation, then, a class X is finite if and only if for some number n, there are *not* more than n elements of X. What does 'more than n elements of X' mean? Here we draw upon Leibniz's subscription to the idea that a number is an aggregate of unities, for instance, that $6 = 1+1+1+1+1+1$ (cf. A VI, 3, 518). Leibniz is explicit about this when he defines 'integer number':

An integer number is a whole [totum] collected from unities. (LH XXXV, 1, 9, f. 7r-v)[36]

Each 'integer number' or natural number is then itself a class, indeed a whole, and it is open to compare other classes to it. It is also clear that Leibniz consciously intends for his definition of an infinite multitude to be understood in terms of com-parisons with numbers taken as wholes composed of unities; for example, he asserts

---

[34] See Ishiguro (1990), Knobloch (1994, 2002), Bassler (1998), Arthur (2008), (2009) and (2013), and Levey (2008). In the present paper it's worth noting that the same crossed-out passage that contains the denial of cardinal equality between the evens and odds, Leibniz begins: "*There is a syncategorematic infinite* or passive power having parts, namely, the possibility of further progress by dividing, multiplying, subtracting, or adding. And *there is a hypercategorematic infinite*, or po-testative infinite, an active power having, as it were, parts eminently but not formally or actually. This infinite is God himself. But *there is not a categorematic infinite* or one actually having infinite parts formally" (GP II, 314–315). Translated by Look and Rutherford in Leibniz (2007, p. 53).

[35] William Heytesbury may have been the first to defend a syncategorematic analysis of 'infinite'; see *sophisma* xviii of his *Sophismata*, in Pironet (1994).

[36] Quoted in Grosholz and Yakira (1998, p. 99).

the existence of infinitely many bodies in this way: 'Bodies are actually infinite, that is, there exist more bodies than there are unities in any given number' (A VI, 4, 1393/Ar 235).

It then remains only to define 'more than n elements of X'. No appeal to Euclid's Axiom that the whole is greater than the part is available here to license the claim that an infinite multitude contains more elements than unities in a natural number, since Leibniz rules out infinite wholes. The most natural route at this point is not via Euclid's Axiom at all but by appeal to the standard of one-one maps in combination with the idea of a natural number as a totality of elements: there are more than n elements of X just in case there is no one-one map from X into the natural number n. This now allows us to state Leibniz's definitions of 'infinite' in canonical terms: X is infinite iff there is no one-one map from X into any natural number.

Assiduous readers will note that Leibniz's definition of 'infinite' is not the same as the definition suggested in our earlier discussion of Galileo. Whereas Galileo's definition considers maps of all the natural numbers into classes—a class is infinite just in case there is such a map, and finite otherwise—Leibniz's definition considers maps from classes into individual natural numbers. Dedekind's definition takes yet another angle, considering maps from classes into themselves: a class is infinite just in case it can be mapped into a proper part of itself. All these definitions are at least conceptually distinct. Intriguingly, Galileo's and Dedekind's definitions turn out to be equivalent, while Leibniz's definition turns out to be different, and weaker, if only barely so. And it is Leibniz's definition that is now the standard definition of infinite in mathematics, while Galileo's and Dedekind's has come to be regarded as a special case called 'Dedekind infinite'. But that is a story for another day.[37]

Acknowledgements My thanks to Katherine Dunlop, Jeffrey McDonough, O. Bradley Bassler, Richard Arthur, Philip Beeley and Norma Goethe for helpful comments, and to audiences at the University of Oxford and Texas A&M University who heard preliminary versions of this material. My thanks also to Michael Detlefsen for the invitation that eventually led to this work, to Eberhard Knobloch whose skepticism about my claims about Galileo in an earlier paper made me to reconsider the issues more carefully, and especially to Norma Goethe, again, who encouraged me to write the present paper.

References

Arthur, Richard. 2008. Leery bedfellows: Leibniz and Newton on the status of infinitesimals. In *Infinitesimals differences: Controversies between Leibniz and his contemporaries*, ed. U. Goldenbaum and D. Jesseph, 7–30. Berlin: Walter de Gruyter.

Arthur, Richard. 2009. Actuals infinitesimals in Leibniz's early thought. In *The philosophy of the young Leibniz (Studia LeibnitianaSonderheft 35)*, ed. M. Kulstad, M. Laerk, and D. Snyder, 11–28. Stuttgart: Franz Steiner Verlag),

Arthur, Richard. 2013. Leibniz's syncategorematic infinitesimals. *Archive for History of Exact Sciences* 67:553–593.

---

[37] See my 'Leibniz's Analysis of Galileo's Paradox,' ms.

Bassler, O. Bradley. 1998. Leibniz on the indefinite as infinite. *The Review of Metaphysics* 15: 849–874.

Benardete, José. 1964. *IInfinity: An essay in metaphysics.* Oxford: Clarendon Press.

Bolzano, Bernard. 1851/1950. *Paradoxes of the infinite* (translated by F. Prihonsky). London: Routledge.

Brown, Gregory. 2005. Leibniz's mathematical argument against a Soul of the World. *British Journal for the History of Philosophy* 13:449–488.

Cantor, Georg. 1915. *Contributions to the founding of the theory of transfinite numbers* (Translated and edited by Philip Jourdain). New York: Dover.

Costabel, Pierre. 1964. La roue d'Aristote et les critiques françaises à l'argument de Galilée. *Revue d'Histoire des Sciences* 17:385–396.

Drabkin, I. E. 1950. Aristotle's wheel: Notes on the history of a paradox. *Osiris* 9:126–198.

Dummett, Michael. 1994. What is mathematics about? In *Mathematics and mind*, ed. Alexander George, 11–26. Oxford: Oxford University Press.

Galilei, Galileo. 1898. *Opere*, Vols. 1–8. ed. Antonio Favaro. Florence: Edizione Nazionale.

Galilei, Galileo. 1974. *Two new sciences, including centers of gravity and force of percussion*, 2nd edn. (Translated and edited with commentary by Stillman Drake). Toronto: Wall and Emerson, Inc.

Grosholz, Emily, and Elhanan Yakira. 1998. *Leibniz's science of the rational. Studia leibnitiana—Sonderhefte 26.* Stuttgart: Franz Steiner Verlag.

Ishiguro, Hidé. 1990. *Leibniz's philosophy of logic and language*, 2nd edn. Cambridge: Cambridge University Press.

Knobloch, Eberhard. 1994. The infinite in Leibniz's mathematics: The historiographical method of comprehension in context. In *Trends in the Historiography of Science*, ed. Kostas Gavroglu et al., 265–278. Dordrecht: Kluwer.

Knobloch, Eberhard. 1999. Galileo and Leibniz: Different approaches to infinity. *Archive for History of Exact Sciences* 54:87–99.

Knobloch, Eberhard. 2002. Leibniz's rigorous foundation of infinitesimal geometry by means of riemannian sums. *Synthese* 133 (1–2): 59–73.

Knobloch, Eberhard. 2011. Galileo and German thinkers: Leibniz. In *Galileo e la scuola galileiana nelle Università del Seicento*, ed. L. Pepe, 127–139. Bologna: Cooperativa Libraria Universitaria Bologna.

Leibniz, G.W. 1849–1863. *Mathematische Schriften von Gottfried Wilhelm Leibniz.* Vols. 1–7 (Ed. C.I. Gerhardt). Berlin: A. Asher; Halle: H.W. Schmidt.

Leibniz, G.W. 1875–1890. *Die Philosophischen Schriften.* Vols. 1–7 (Ed. C.I. Gerhardt). Berlin: Weidmannsche Buchhandlung.

Leibniz, G.W. 1923–1999. *Samtliche Schriften un Briefe. Philosophische Schriften.* Series VI. Vols. 1–4. Berlin: Akademie-Verlag.

Leibniz, G.W. 1948. *Texts inédits d'apres les manuscits de la bibliothéque provinciale de Hanover.* Vols. 1–2. Edited by Gaston Grua. Paris: P.U.F.

Leibniz, G.W. 2001. *Labyrinth of the continuum: writings on the continuum problem, 1672–1686.* Translated and edited with commentary by Richard T. W. Arthur. New Haven: Yale University Press.

Leibniz, G.W. 2007. *The Leibniz-Des Bosses correspondence.* Translated and edited with commentary by Brandon Look and Donald Rutherford. New Haven: Yale University Press.

Levey, Samuel. 1998. Leibniz on mathematics and the actually infinite division of matter. *The Philosophical Review* 107 (1): 49–96.

Levey, Samuel. 1999. Leibniz's constructivism and infinitely folded matter. In *New essays on the rationalists,* ed. Rocco J. Gennaro and Charles Huenemann, 134–162. New York: Oxford University Press.

Levey, Samuel. 2003. The interval of motion in Leibniz's *Pacidius Philalethi. Noûs* 37:317–416.

Levey, Samuel. 2008. Archimedes, infinitesimals and the law of continuity: On leibniz's fictionalism. In *Infinitesimals differences: Controversies between Leibniz and his contemporaries,* ed. U. Goldenbaum and D. Jesseph, 107–134. Berlin: Walter de Gruyter.

Levey, Samuel. 2011. On two theories of substance in Leibniz: Critical notice of Daniel Garber, *Leibniz: Body, Substance, Monad. The Philosophical Review* 120:285–319.

Lison, Elad. 2006/2007. The philosophical assumptions underlying Leibniz's use of the diagonal paradox in 1672. *Studia Leibnitiana* 38/39:197–208.

Mancosu, Paulo. 1996. *Philosophy of mathematics and mathematical practice in the seventeenth century.* New York: Oxford University Press.

Mancosu, Paulo. 2009. Measuring the size of infinite collections of natural numbers: Was cantor's theory of infinite numbers inevitable? *Review of Symbolic Logic* 2:612–646.

Netz, R., K. Saito, and N. Techenetska. 2001–2002. A new reading of method proposition 14 from the Archimedes Palimpsest (Parts 1 and 2). *Sciamus* 2–3:9–29 and 109–125.

Nicholas of Cusa. 1440/1985. *On learned ignorance.* In *Nicholas of Cusa on Learned Ignorance: A Translation and Appraisal of De Docta Ignorantia,* ed. Jasper Hopkins, 2nd ed. Minneapolis: Arthur J. Banning Press.

Parker, Matthew W. 2009. Philosophical method and galileo's Paradox of infinity. In *New perspectives on mathematical practices,* ed. Bart Van Kerkhove, 76–113. Singapore: World Scientific.

Pironet, Fabienne, ed. 1994. *Guillaume Heytesbury, Sophismata asinina. Une introduction aux disputes médiévales. Présentation, édition critique et analyse.* , ed. Bart Van Kerkhove, 76–113. Collection Sic et Non; Paris: Vrin.

Russell, Bertrand. 1903. *The principles of mathematics.* Cambridge: Cambridge University Press.

Russell, Bertrand. 1914. *Our knowledge of the external world.* Chicago: Open Court Publishing Co.

Skyrms, Brian. 1983. Zeno's Paradox of Measure. In *Physics, philosophy and psychoanalysis,* ed. R. S. Cohen and L. Lauden, 223–254. Dordrecht: D. Reidel Publishing Company.

Leibniz on The Elimination of Infinitesimals

Douglas M. Jesseph

There can scarcely be any question that the most important of Leibniz's many mathematical contributions was his development of the *calculus differentialis*. Although his work in such fields as algebra, series summations, combinatorics, determinant theory, and other areas can be seen (retrospectively at least) as original and even groundbreaking, his efforts in these disciplines were not generally made public his day. In contrast, historians of mathematics routinely speak of a "Leibnizian tradition" in analysis that traces back to a series of published papers from the 1680s and constitutes a clear advance in European mathematics. Yet, although the infinitesimal methods Leibniz introduced with his calculus were powerful new techniques opening a vast field of new results, these methods raised significant conceptual and methodological issues. The specific worry voiced by several of Leibniz's mathematical contemporaries was that these new methods were fundamentally unrigorous, so that they not only offended against criteria of mathematical intelligibility, but were also liable to lead to error.

1 The Traditional Standard of Mathematical Rigor

As traditionally conceived, a mathematical demonstration must begin with first principles that are transparently true and proceed deductively to the derivation of theorems. The first principles take the form of definitions, axioms (i.e., general principles applicable to any science whatever) and postulates, or principles specific to a given science. Aristotle famously declared that the first principles upon which demonstrative sciences depend must themselves be indemonstrable and known by a

D. M. Jesseph (✉)
Department of Philosophy, South Florida University, Tampa, FL, USA
e-mail: djesseph@usf.edu

© Springer Netherlands 2015
N. B. Goethe et al. (eds.), *G.W. Leibniz, Interrelations between Mathematics and Philosophy,* Archimedes 41, DOI 10.1007/978-94-017-9664-4_9

kind of immediate, non-inferential understanding (*Posterior Analytics* 72ᵇ 19–21).[1] Of course, not just any first principles are permissible in mathematics, for the obvious reason that the security of the derived theorems can be no greater than that of the principles from which the theorems are derived. The result is Aristotle's famous requirement that the first principles of demonstration must be "true, and primitive and immediate and more familiar than and prior to and explanatory of the conclusion" (*Posterior Analytics* I.2, 70ᵇ 21–23).

This general model of rigorous mathematical demonstration persisted well beyond classical Greek mathematics. Indeed, seventeenth-century philosophers and mathematicians were practically unanimous in their acceptance of the traditional ideal. Philosophically-minded mathematicians who disagreed about nearly everything else could at least be united in the opinion that mathematical demonstrations must take into consideration only things that are clearly conceived and proceed deductively by truth-preserving inferences from definitions and axioms that cannot be doubted. Thus, such diverse figures as Descartes, Pascal, Hobbes, Wallis, and Barrow (who could agree on essentially nothing else) were unanimous in holding that the objects of mathematical investigation must be the sorts of things that are evident to reason.[2]

The requirement that rigorous demonstrations must be concerned only with things that can be clearly conceived had an obvious corollary in ancient mathematics, namely that infinitary methods are to be banished as essentially unrigorous. The basis for this banishment is obvious. Aside from failing the test of clear intelligibility, the infinite famously threatens paradox, and classical mathematics is thus developed with a very strong finitistic bias. As a result, the incomprehensibility of an (actual as opposed to potential) infinite was taken to be perfectly in keeping with proper mathematical theory and practice. Indeed, Aristotle declared that "In point of fact [mathematicians] do not need the infinite and do not use it" (Physics IV.7, 207b 30), and this attitude persisted until well into the seventeenth century.

Although the traditional account of rigor forbids the appeal to any actual infinite, it did license the use of a potential infinite. This familiar distinction allows the assertion that, for any given positive magnitude, a smaller one may still be taken; but it does not permit one to conclude that there exists a magnitude smaller than any given positive magnitude. The "exhaustion"proofs of classical Greek geometry make use of this idea of a merely potential infinite, by showing that the difference between a sought result and a sequence of approximations can be made less than any given magnitude.

---

[1] The Aristotelian account of demonstrative knowledge is summarized and analyzed in Hankinson (1995). Aristotle's treatment of first principles is aptly summarized in Heath's commentary to the Euclidean *Elements*, (Euclid [1925] 1956, 1: 117–123).

[2] On Descartes' conception of rigor, see Bos (2001, Chap. 15); Pascal's conception of rigorous proof is articulated in his essay "L'esprit géométrique" in Pascal (1963, pp. 348–359); Hobbes's account of demonstration is summarized in the first of his *Six Lessons* (1656); Wallis's conception of rigor is outlined in the first three chapters of his *Mathesis Universalis* in Wallis (1693-1699, 1: 17–24); Barrow's approach is summarized in the fourth of his *Lectiones Mathematicae* (Barrow 1860, 1: 63–76).

   The canonical formulation of exhaustion proofs was set out by Archimedes, who proceeded by constructing sequences of approximations that systematically converge on the sought result.[3] The logical form of an exhaustion proof can be set out as follows. We seek a magnitude, which we can call μ. To begin, we construct a sequence a_n of rectilinear approximations less than μ such that

$$a_0 < a_1 < \ldots a_n < \mu, \text{ where } |\mu - a_{k+1}| < \tfrac{1}{2}|\mu - a_k|.$$

Next, we construct a sequence A_n of rectilinear approximations greater than μ such that

$$A_0 > A_1 > \ldots > A_n, \text{ where } \tfrac{1}{2}|A_k - \mu| > |A_{k+1} - \mu|.$$

The sequence a_n gives successively better approximations to the sought value "from below", as when a sequence of inscribed polygons approximate the area of a curvilinear figure; and the sequence A_n gives ever-better approximations to the sought value "from above", as when a sequence of circumscribed polygons approximate the area of a curvilinear figure. When both sequences satisfy the condition that the difference between each successive approximation and the sought value is reduced by more than half as we move from term to term of the sequence, then they both "compress" the desired value between two sequences that converge to a common limit. The proof can then be completed with a pair of *reductio ad absurdum* arguments, showing that the result could be neither greater nor less than the common limit μ, to which the compressing sequences converge.

   The method of proof by exhaustion was the standard for rigorous demonstration in the seventeenth century, but it was also regarded as cumbersome, prolix, and incapable of generality (Whiteside 1960–1962, pp. 330–335). In many cases exhaustion proofs were used to confirm or establish results that had been achieved by other methods, but exhaustion itself seemed poorly suited to the task of uncovering new results. A central drawback in the method lies in the inherent difficulty of constructing sequences of rectilinear approximations that can be systematically improved and shown to converge. In all but the simplest cases, generating such sequences is a difficult task, and it becomes significantly more difficult as the curves get more complex. In his *Géométrie* of 1637, Descartes argued that a class of curves much broader than that countenanced by ancient geometers should be recognized as truly geometrical, on the grounds these could be defined through criteria that enabled "precise and exact" determination.[4] But extending the exhaustion method to cover these more complex curves makes the application of the technique all but hopeless: exhaustion finds a straightforward application to the case of simple "compass and rule" constructions, and the method generalizes fairly easily to conic sections, but any further extension of the method faces serious technical obstacles. Small wonder, then, that seventeenth century mathematicians began to rely upon infinitesimal

---

[3] On the Archimedean method of exhaustion, see Dijksterhuis (1987, pp. 130–134).

[4] On Descartes and the question of what counts as properly geometrical, see Bos (2001).

considerations in the solution of important outstanding problems, even as they paid lip service to the classical ideal of rigorous demonstration.

In fact, we can see interesting evidence of this in Leibniz's writings from the 1670s, where he sought a general method for solving problems of tangency and quadrature that will retain the "apagogical" nature of exhaustion while serving as a method of discovery. In a brief note from 1674 entitled "On the use of inscribed and circumscribed [figures] not only for demonstration, but also for discovery" he declared that "The apagogic method, by for instance inscribed and circumscribed [figures] has thus far been used by geometers only for the purpose of demonstrating things that have been discovered by other means. I began to consider whether or not this very general method might not also be a principle of discovery, so that it comprehends everything else, since everything discovered by other methods can be demonstrated by this one" (A VII, 5, 113). The search for a general method of this sort ultimately led Leibniz to the infinitesimal calculus, which we can now consider.

2   Leibnizian Calculus and the Infinitesimal

In 1684 Leibniz published his first paper on the calculus, which bore the title "A New Method for Maxima and Minima, as well as Tangents, which is not Impeded by Fractional or Irrational Quantities, and a Remarkable Type of Calculus for Them" (GM V, 220–226).[5] As the title suggests, this Leibnizian paper promises a new approach to problems of maxima and minima, one that will overcome the difficulties that had plagued earlier attempts to study complex curves involving fractional or irrational powers, and to do all of this with a simple calculus or routine procedure that can be applied systematically to all manner of problems.

The history of the calculus is well enough known that I do not need to repeat it in any detail, but it is important to draw attention to the conceptual difficulties posed by the infinitesimal. The great power of the Leibnizian calculus is that it reduces the inverse problems of constructing tangents and determining areas to a simple algorithmic procedure based on the comparison of increments between quantities x and y in an "analytic" equation representing a curve. Leibniz's general approach is to examine the behavior of such a curve when infinitesimal increments dx and dy are introduced. Replacing x and y with $(x+dx)$ and $(y+dy)$ yields a new equation that can be manipulated by treating the infinitesimals dx and dy as positive quantities obeying the ordinary rules for such operations as addition, multiplication, subtraction, division, or extraction of roots. Yet, at need, these infinitesimals can also be discarded by taking them to be effectively zero. The result is a method for constructing tangents, determining quadratures, finding arc length, and solving

---

[5] The key papers on Leibniz's calculus are collected in Leibniz (1995). Roero (2005) has an overview of Leibniz's early publications on the calculus. See Parmentier (1995) for an account of Leibniz's calculus in the context of his mathematical "optimism."

a wide variety of other problems, regardless of the complexity of the curves studied or the nature of the equations that define them.

It should be clear from the start that there is something rather puzzling going on here. The differential increments dx and dy appear to have contradictory properties, depending on the needs of the moment. Whenever we need to subtract, divide, or raise a variable to a power, the increments are assumed to be positive; yet when we want to discard terms containing such quantities, they are tacitly assumed equal to zero. This puzzling duality of the infinitesimal sets the stage for a "logical" objection to the *calculus differentialis*, namely that it depends upon inconsistent assumptions. A second kind of criticism is a "conceptual" or "metaphysical" complaint to the effect that infinitesimals are not clearly conceivable or definable, thereby violating the standard criterion of rigorous demonstration.

Proponents of the Leibnizian calculus sought by various approaches to allay these concerns; some announced that differential increments were quantities greater than zero, yet less than any finite magnitude. Others declared an infinitesimal to be a magnitude bearing the same ratio to a finite magnitude that a finite magnitude bears to an infinite magnitude[6]. The first characterization of infinitesimals seeks to situate the infinitesimal between something and nothing, thereby giving it a sort of "hybrid" existence that can be taken as alternately a positive magnitude or zero. This second way of thinking about infinitesimals attempts to capture their apparently inconsistent properties: as ratios of positive magnitudes, infinitesimals remain greater than nothing; but as ratios of finite to infinite, they are less than any positive quantity (since positive quantities can always be expressed as ratios of finite magnitudes). Aside from the conceptual difficulties posed by infinitesimal increments such as dy or dx, the Leibnizian calculus uses higher-order differentials such as dy^2 or dx^2. As products of two infinitesimal quantities, second-order differentials appear to be greater than zero, yet less than any infinitesimal of the first order, and indeed this is how they were typically characterized both by Leibniz and his associates.

Although the new calculus opened vast new areas of mathematical research and offered a very general means to solve problems that had previously been attacked on a piecemeal basis, its procedures offended quite grossly against the traditional standard of mathematical rigor. Whatever virtues one might claim for them in the solution of problems involving tangency, quadrature, arc-length determination, maxima, or minima, the fact remains that infinitesimal magnitudes are surely not the sort of thing anyone can claim to conceive clearly. Magnitudes that hover between something and nothing are surely difficult to conceive; and the mere fact that we have a notation involving terms 'dx' or 'dy' doesn't mean that we have anything like a guarantee that there really are things denoted by or answering to our notation. The result, not surprisingly, is that the new calculus faced objections from "traditionalist" opponents, who charged Leibniz and his associates with violating obvious criteria of geometric rigor.

---

[6] The *locus classicus* for such characterizations of the infinitesimal is L'Hôpital (1696).

These objections have been discussed at length in the recent literature on Leibniz, and I do not need to go over them in any detail.[7] It suffices for my purposes to mention the objections raised Michel Rolle in 1700. In league with the Abbé Jean Gallois and the Abbé Thomas Gouye, Rolle made the case against the "new system of the infinite" in a session of the *Académie Royale des Sciences* in July of 1700. In doing so, Rolle endorsed the classical ideal of rigorous demonstration with the remark:

> We have always regarded geometry as an exact science, and also as the source of the exactness that is spread throughout all the other parts of mathematics. We see among its principles only true axioms: and all theorems and all the problems proposed here are either solidly demonstrated or capable of solid demonstration. And if it should happen that any false or less certain principles should slip in, they must at once be banished from this science.
> But it seems that this character of exactitude no longer reigns in geometry, ever since we became entangled in the new system of the infinitely small. (Rolle 1703, p. 312)

Rolle extended this criticism of the intelligibility of the foundational concepts of the calculus with objections intended to show that it actually produced errors when applied to specific cases.[8] The defense of the new methods was not taken up by Leibniz himself (who would have been overjoyed to return to Paris but was constrained by circumstances to reside in Hanover). Instead, Pierre Varignon served as the chief defender of the calculus, though he remained in steady epistolary contact with Leibniz himself.

Leibniz's reply to such criticisms was to insist that, when properly understood, his calculus does not require truly infinitesimal magnitudes. The general strategy he pursued was to propose that any apparent commitment to the reality of infinitesimal magnitudes could be sanitized by regarding the infinitesimal as a sort of "useful fiction" that is harmlessly introduced into derivations without undermining the rigor of the underlying reasoning. In a letter published in the *Journal des Savants* in 1701 he declared:

> [T]here is no need to take the infinite here rigorously, but only as when we say in optics that the rays of the sun come from a point infinitely distant, and thus are regarded as parallel. And when there are more degrees of infinity, or of infinitely small, it is as the sphere of the earth is regarded as a point in respect to the distance of the sphere of the fixed stars, and a ball which we hold in the hand is also a point in comparison with the semidiameter of the sphere of the earth. And then the distance to the fixed stars is infinitely infinite or an infinity of infinities in relation to the diameter of the ball. For in place of the infinite or the infinitely small we can take quantities as great or as small as necessary in order that the error will be less than any given error. In this way we only differ from the style of Archimedes in the expressions, which are more direct in our method and better adapted to the art of discovery (GM IV, 95–96).

Leibniz's apparent admission that the infinite need not be taken "rigorously" caused some of his associates and defenders of the calculus—notably Jean Bernoulli, Varignon, and the Marquis de L'Hôpital– to request a clarification of his views.

---

[7] See Goldenbaum and Jesseph (2008) for an overview of objections to the Leibnizian infinitesimal raised by his contemporaries. Mancosu (1996, Chap. 6) deals with some of these objections as well.

[8] See Mancosu (1996, pp. 168–176) for more on Rolle's objections.

The result was a very important letter from February of 1702 in which Leibniz set out his "fictionalist" reading of the infinitesimal, noting that although infinitely small magnitudes cannot be real, nevertheless "everything in geometry and even in nature takes place as if they were perfect realities" (GM III, 93). As Leibniz recounted the events in a late letter to Pierre Gangicourt, "[w]hen our friends were disputing in France with the Abbé Gallois, father Gouyé and others, I told them that I did not believe at all that there were actually infinite or actually infinitesimal quantities; the latter, like the imaginary roots of algebra ($\sqrt{-1}$) were only fictions, which however could be used for the sake of brevity or in order to speak universally.... But as the Marquis de L'Hôpital thought that by this I should betray the cause, they asked me to say nothing about it, except what I had already communicated in the Leipzig *Acta*" (Leibniz 1768 3: 500–501).

It is tolerably clear that Leibniz was a committed fictionalist about infinitesimal magnitudes by 1702. His view of infinitesimals as "useful fictions" seems to have taken shape in the mid 1690s, although there are certainly traces of it as early as the 1670s, and a forthright statement of the fictionalist position seems to have come from Leibniz's pen only in the aftermath of the dispute in the *Académie des Sciences* over the foundations of the calculus.[9] The question to consider is whether fictionalism can be squared with traditional notions of rigor: i.e., whether there can be a rigorous theory of fictions.

3   Fictionalism and the Question of Rigor

Philosophically, the recourse to fictionalism poses something of a problem for the Leibnizian account of rigorous demonstration. Leibniz never voiced serious reservations about the traditional standard of mathematical rigor, although he did remark "excess of scruple" might hinder the art of discovery and deny us its fruits.[10] Indeed, Leibniz's own account of demonstration seems to go further than the traditional demands that the first principles of mathematics be clearly conceived, since he held

---

[9] The literature on the Leibnizian understanding of infinitesimals is extensive. The most complete study is Bos (1974), which distinguishes two Leibnizian approaches to the infinitesimal: one relying upon exhaustion, the other involving the Leibnizian "law of continuity", intended to provide a theoretical basis for Leibniz's fictionalism. Ishiguro (1990, Chap. 5) reads Leibnizian fictionalism as adopting a "syncategorematic" view in which infinitesimal terms do not denote and play the role of logical fictions. This interpretation has been endorsed and extended by Arthur (2008), Knobloch (1994, 2002), and Levey (2008). Precisely when Leibniz committed to the fictionalist approach is a subject of some scholarly disagreement. Levey (2008, p. 107) declares that Leibniz "abandoned any ontology of actual infinitesimals" in the 1670s, and both Arthur and Knobloch accept variants of this timeline. I have argued elsewhere (Jesseph 1998) that Leibniz's correspondence with Wallis and Jean Bernoulli in the 1690s betrays serious reservations about the reality of infinitesimals (in opposition to Bernoulli) but also a disinclination to take them as "nothing" (contrary to Wallis). Nothing I say here depends upon settling this specific issue.

[10] This remark appears in Leibniz's response to the criticisms of the calculus voiced by Bernard Nieuwentijt (GM V, 322).

that all demonstrations must ultimately terminate in identities, and even traditional axioms or postulates could ultimately be demonstrated. This rather extreme view is expressed in a late essay on the "Metaphysical basis of mathematics", where Leibniz claimed that "demonstrations are finally resolvable into two kinds of indemonstrables: *definitions* or ideas, and *primitive propositions* or identities, such as *A* is *A*, anything whatever is equal to itself, and a great many others of this kind" (GM VII, 20). Or, as Leibniz put the matter in a somewhat different context "[Geometers] are agreed upon axioms and postulates, upon whose truth the rest [of geometry] depends. We accept these, both because they satisfy the mind immediately and because they are proved by countless experiences; nevertheless, it would be an aid to the perfection of the science to prove them. This was attempted of old for certain axioms by Apollonius and Proclus and recently by Roberval... I am convinced that the demonstration of the axioms is of great assistance to true analysis or the art of discovery" (GP IV, 354).

Given this (or any other reasonable conception of rigorous demonstration), simply declaring infinitesimals to be fictional does not, by itself, do anything to show that inferences based upon such fictions are truly rigorous. This point can be easily enough illustrated by an example. Suppose that, in the interest of advancing number theory, I define the number p to be the largest prime number, i.e. the greatest finite integer with only trivial divisors. Given p, I have a quick and easy solution to the twin prime conjecture: since there are only finitely many primes up to and including p, there are only finitely many primes of the form n, $(n+2)$. Moving along, I find a simple (albeit non-constructive) counterexample to the Goldbach conjecture: since p is the greatest prime, $2p+2$ is an even number that cannot be expressed as a sum of two primes. I have now settled two famous outstanding conjectures in number theory with elementary proofs, so the prospects for tremendous advances in this field must seem very bright indeed. If someone objects to the viability of these results, I will reply that p is simply a fiction: I do not really believe that there is a greatest prime, but I find the supposition useful for simplifying number theory and proving new results. And who can doubt that this fictional prime p has greatly simplified and expanded number theory?

The obvious problem here is that assuming the existence of p is to assume something that is demonstrably inconsistent. And, in general, the recourse to fictionalism is insufficient on its own to make a demonstration employing such fictions truly rigorous or convincing. To vindicate fictionalism, the introduction of a fictional entity must be accompanied by some kind of guarantee to the effect that indulging in the fiction does not require anything inconsistent and will never lead to error. Thus, a central task for Leibniz's fictionalism about infinitesimal magnitudes is to show that their introduction is essentially harmless in the sense that it does not yield a contradiction.

There seem to be two strategies that might reconcile Leibniz requirements for rigorous demonstration with his doctrine of the fictionality of the infinitesimal. The first of these is a "syntactic" or "proof theoretic" strategy that would attempt to show how any derivation that invokes infinitesimals could, in principle, be re-written in the form of an Archimedean exhaustion proof. Pursuing this strategy would involve showing that any appearance of an infinitesimal quantity like dx in the course of a demonstration could be replaced by a more complex expression involving sequences of approximations that compress or converge to the sought result. In the end, this approach would take the fictionality of the infinitesimal as amounting to the claim that a symbol like 'dx' is simply a placeholder for a much more elaborate line of reasoning that makes reference only to finite differences of finite quantities. Put somewhat more formally, this strategy would seek to show that for any axioms O for "ordinary" analysis and axioms I governing infinitesimals, whenever $O \cup I \vdash \varphi$ and 'φ' does not contain infinitesimal terms, we have a means to construct a proof of from O alone, that is, $O \vdash \varphi$.

A second strategy would be a "semantic" or "model-theoretic" approach that seeks to show that, even if reference to infinitesimals cannot be eliminated from analysis, we can still have confidence that the use of infinitesimal magnitudes will never lead from truth to falsehood. This line of justification would try to show that any formula that is provable using infinitesimals (but does not itself contain infinitesimal terms) is nevertheless a semantic consequence of standard axioms of geometry. Here, the idea is that, taking O and I as axioms for "ordinary" and "infinitesimal" analysis, in any case where $O \cup I \vdash \varphi$, we are guaranteed that $O \vDash \varphi$. To make this sort of strategy work, Leibniz would have to offer something like a proof of the principle that adding infinitesimals to the standard geometry yields a model-theoretic conservative extension of standard geometry.

I do not wish to indulge in the anachronism of attributing to Leibniz a clear understanding of our proof-theoretic and model-theoretic concepts, and I certainly doubt that these two sorts of strategies were clearly separated in his mind. I do, however, think that we can see glimpses of them in two very different ways that Leibniz tried to accommodate the infinitesimal within the framework of classical geometry. The first of these appears in Leibniz's *Arithmetical Quadrature of the Circle, Ellipse, and Hyperbola*, written in 1675–6; the second is in his "Note on the Justification of Infinitesimal Calculus by that of Ordinary Algebra" which he forwarded to Varignon in 1702 in the context of the dispute in the Académie. I now turn to a brief consideration of these texts.

3.1 Leibniz's Arithmetical Quadrature of the Circle

The pace of Leibniz's mathematical research in the 1670s is really quite remarkable. He arrived in Paris in 1672 on a diplomatic mission for the elector of Mainz and had essentially no mathematical training.[11] After a few years, he had mastered the cutting-edge mathematics of his day, and around 1675 he produced the longest mathematical treatise he ever wrote, *On the Arithmetical Quadrature of the Circle, the Ellipse, and the Hyperbola*. The work was never published in Leibniz's

---

[11] Hofmann (1974, p. 2) aptly characterizes the state of Leibniz's mathematical knowledge upon his arrival in Paris as "deplorable."

lifetime and its first appearance in print came in 1993 in an edition prepared by Eberhard Knobloch. The remarkable feature of this treatise is that in it Leibniz offers a rigorous foundation for the theory of infinitesimal magnitudes, at least in the application to a class of problems involving conic sections. Others have discussed the *Arithmetical Quadrature* at length,[12] and my treatment of the issues will be correspondingly brief. It is nevertheless worthwhile to consider the general approach that Leibniz undertook to show that infinitesimal reasoning can be replaced by strict Archimedean exhaustion proofs in a variety of contexts.

We should begin by noting that even as early as the mid-1670s Leibniz was prepared to hedge his bets on the reality of infinitesimals and endorse something very much like his later fictionalism. As he put it in the *Arithmetical Quadrature*, "Nor does it matter whether there are such quantities in nature, for it suffices that they are introduced as fictions, since they allow the abbreviations of speech and thought in the discovery as well as demonstration" (Leibniz 1993, p. 69). This fictionalism is grounded in Leibniz's argument to the effect that any proof involving infinitesimal magnitudes can be re-written in the form of a classical exhaustion procedure.

The fundamental theorem in Leibniz's attempt to show the eliminability of infinitesimals is the sixth proposition of the *Arithmetical Quadrature*, which he characterizes as "most thorny" and one whose "reading may be omitted if one does not want the greatest rigor" in the demonstration of other results (Leibniz 1993, p. 28). Leibniz's reasoning relates to the construction in the figure (see Fig. 1). Regrettably for contemporary readers, his notational conventions are prone to engender confusion: the y axis is horizontal across the top and the x axis is vertical toward the bottom, while the subscripts are placed to the left rather than the right in the labels.

Leibniz argues as follows: Let $A_1C_2C_3C$ etc. be a circular arc. We then construct the points of intersection of the tangents at points $_iC$ with the horizontal axis, yielding $_1T, _2T, _3T$ etc. Next, we define an auxiliary curve $_1D, _2D, _3D$, etc. by pairing the ordinates $_1B, _2B, _3B$, etc. corresponding to the points $_1C, _2C, _3C$, etc. in the circular arc with the abscissae $_1T, _2T, _3T$, etc. Then we take the secants joining successive points $_1C, _2C, _3C$, etc. in our division of the circle and extend them to cut the horizontal axis at the points $_1M, _2M, _3M$, etc. Finally, we take perpendiculars from these points and pair them with the ordinates $_1B, _2B, _3B$, etc. to generate another curve through the points $_1N, _2N, _3N$, etc. The two new curves are then used to define curvilinear areas whose exact measure is determined by rectilinear approximations in the style of Archimedes, including the complex *reductio* proofs. The original case of the circle can then be extended by considering other conic sections (notably the ellipse and hyperbola), that admit of an analogous treatment.

Knobloch characterizes this approach as equivalent to showing "the integrability of a huge class of functions by means of Riemannian sums which depend on intermediate values of the partial integration intervals" (Knobloch 2002, p. 63). Arthur (2008, p. 24) deems the method "extremely general and rigorous", and his opinion is shared by Levey, who holds that Leibniz's "technical accomplishments in quad-

---

[12] See Arthur (2008), Knobloch (2002), Levey (2008) for accounts of the treatise and its role in Leibniz's mathematics.

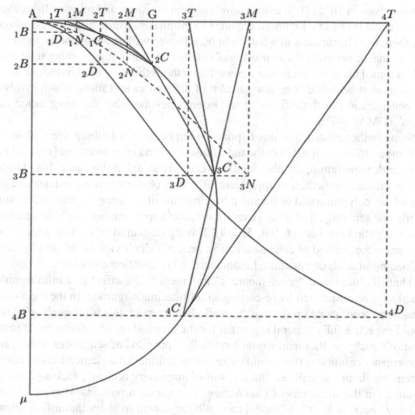

Fig. 1 Based on Leibniz, De Quadratura Arithmetica, Prop. 6

ratures far outstrip the original reaches of the method of exhaustion" (2008, p. 119). Although I agree that Leibniz's procedure can show the eliminability of infinitesimals from a large class of problems, it nevertheless falls short of a completely general method that could apply to any curve representable by an analytic equation. The problem is that the construction of the auxiliary curves requires that we have a tangent construction that will apply to the original curve. This is readily available in the case of the circle, and tangents to conic sections and other well-behaved curves are also constructible with classical methods. However, one great strength of the infinitesimal calculus is that it enables algorithmic solutions to problems of tangency for a wide variety of curves; yet the procedure in the *Arithmetical Quadrature* could only be made fully general if we already had a solution to the general problem of tangent construction. In fact, Leibniz's approach here suffers from the kind of limitations that seventeenth century mathematicians routinely found wanting in the method of exhaustion: it is perfectly rigorous, but limited in its scope of application.

Indeed, the attraction of the infinitesimal calculus is precisely that its algorithmic procedure is not (to use Leibniz's phrase) "impeded by fractional or irrational quantities" and can be applied more generally. My conclusion is that, although Leibniz's investigations in 1675–1676 could show how conic sections and other well-behaved curves could be handled without recourse to infinitesimals, he himself understood that there were limitations to what could be achieved with these methods. Indeed, I suspect that he set aside the *Arithmetical Quadrature* without publishing it because he had turned his attention to more powerful methods that he would introduce in the 1680s in what he called "our new calculus of differences and sums, which involves the consideration of the infinite", and "extends beyond what the imagination can attain" (GM V, 307).

Some further evidence of this hypothesis can be seen in Leibniz's remark in his November, 1674 fragment "On the use of inscribed and circumscribed [figures] not only for demonstration, but also for discovery." He stated in this piece that in order for the exhaustion method to apply generally "it is obvious that we require that the area of any polygon could be obtained by some rule or summary procedure, or more briefly and generally, that some general method of approximation could be obtained by an analytic calculus" (A VII, 5, 114). Having despaired of finding a means of converting the method of exhaustion into a general calculus applicable to any curve defined by an analytic equation, Leibniz turned his attention elsewhere.

Thus, the attempted "proof-theoretic" argument to the effect that infinitesimals can always be eliminated by re-casting an infinitesimal argument in the form of an exhaustion proof came up short of the mark. In retrospect, this is precisely what one would expect: a fully general argument for the elimination of infinitesimals in the style of Cauchy or Riemann would require the apparatus of sequences, series, and convergence conditions that would make talk of infinitesimal elements dissolve into statements about inequalities. But this sort of machinery is clearly lacking from the argument in the *Arithmetical Quadrature*. The question remains whether Leibniz had any better luck with a "model theoretic" argument to show that infinitesimals would never lead from truth to falsehood, and I now turn to a consideration of an example of this approach.

3.2   The Justification of the Infinitesimal Calculus by that of Ordinary Algebra

Leibniz's important letter to Varignon of February, 1702 includes an enclosure that purports to justify the infinitesimal calculus by means of the "calculus" of what Leibniz termed "ordinary algebra." The idea here is that the algebra of ordinary finite quantities offers a method that can be used to show that appeal to infinitesimals never leads us from truths to falsehood. So, rather than attempt to show that any proof that employs infinitesimal can be rewritten in the classical form, Leibniz undertook to show that the introduction of infinitesimals is ultimately harmless and could be underwritten by general algebraic principles. I reproduce Leibniz's reasoning in its entirety (See Fig. 2):

> Let two straight lines *AX* and *EY* meet at *C*, and from points *E* and *Y* drop *EA* and *YX* perpendicular to the straight line *AX*. Call *AC c* and *AE e*; call *AX x* and *XY y* (as in Fig. 2.

Fig. 2 Leibniz's Justification
of Infinitesimals by "ordinary
algebra" Enclosed in a letter
to Varignon 2 February, 1702
(GM IV, 102–104)

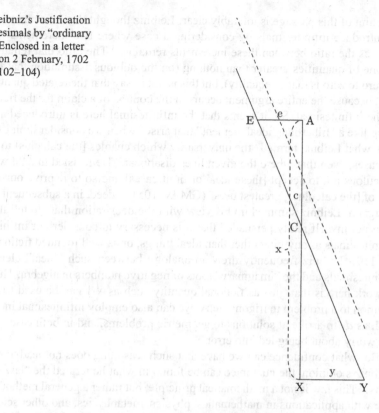

Then since the triangles CAE and CXY are similar, it follows that $\frac{(x-c)}{y} = \frac{c}{e}$. Consequently, if the straight line EY approaches the point A ever more closely, always preserving the same angle at the variable point C, the straight lines c and e will obviously diminish steadily, yet the ratio of c to e will remain constant. Here we assume that the ratio is other than that of equality and that the given angle is less than half a right angle.

Let us now consider the case where the straight line EY reaches A itself; it is manifest that the points C and E will fall on A, that the straight lines AC and AE, or c and e, will vanish, and that the proportion or equation $\frac{(x-c)}{y} = \frac{c}{e}$ will become $\frac{x}{y} = \frac{c}{e}$, supposing that this case falls under the general rule $\frac{(x-c)}{y} = \frac{c}{e}$. Nevertheless, c and e will not be absolutely nothing,

since they still preserve the ratio of CX to XY, or the ratio between the entire radius and the tangent of the angle at C, the angle which we assumed to remain always the same as EY approached the point A. For if c and e were nothing in an absolute sense in this calculation,

then in the case when the points C, E, and A coincide, c and e would be equal, since one zero equals another, and the equation or proportion $\frac{(x-c)}{y} = \frac{c}{e}$ would become $\frac{x}{y} = \frac{0}{0} = 1$; that is, $x = y$, which is an absurdity, since we assumed that the angle is not half of a right angle. Hence c and e are not taken for zeros in this algebraic calculus, except comparatively in relation to x and y; but c and e still have an algebraic relation to each other. And so they are treated as infinitesimals, exactly as are the elements which our differential calculus recognizes in the ordinates of curves for momentary increments and decrements. Thus we find in the calculations of ordinary algebra traces of the transcendent differential calculus and the same peculiarities about which some scholars have scruples. (GM IV, 104–105)

The aim of this passage is tolerably clear: Leibniz thought that he could show how to introduce infinitesimals by considering a case where two increments "vanish" even as the ratio between these increments remains.[13] The ratio that remains must be one of quantities greater than nothing (on the dubious assumption that the ratio of zero to zero is that of equality), but this is not to say that the related quantities are real, because the entire argument occurs in the context of a claim for the fictionality of the infinitesimal. So, it seems that the infinitesimal here is introduced as something like a Hilbertian "ideal element" that arises when we consider limit cases and seek what Leibniz termed "the universality which enables [the calculus] to include all cases, even that where the given lines disappear." He insisted that "it would be ridiculous not to accept [these ideal or limit cases] and so to deprive ourselves of one of [the calculus'] greatest uses" (GM IV, 105). Indeed, in a subsequent letter to Varignon, Leibniz summed up his view with the declaration that "to tell the truth, I am not myself really persuaded that it is necessary to consider our infinities and infinitesimals as anything other than ideal things, or as well founded fictions" (GM IV, 110). Leibniz frequently drew an analogy between such "ideal" elements as infinitesimals and the "imaginary" roots of negative numbers in algebra. The notion at work here is that, just as fictional quantity such as $\sqrt{-1}$ can be used to find the solution to a problem in trigonometry, we can also employ infinitesimal increments such as dx to arrive at solutions to geometric problems, and in both cases we need not worry about being led into error.

But what confidence can we have that such reasoning does not lead to error? In Leibniz's opinion, the guarantee can be found in what he termed the "law of continuity." This law is not a mathematical principle, but rather a general methodological rule with applications in mathematics, physics, metaphysics, and other sciences. In a 1713 letter to Christian Wolff concerning the "science of infinity" and published in the Leipzig *Acta eruditorum*, Leibniz referred to the law as "first proposed in Pierre Bayle's *Nouvelles de la République des Lettres*, and applied to the laws of motion" (GM V, 385). The law states that "with respect to continuous things, one can treat an external extreme as if it were internal, so that the last case, even if it is of wholly diverse nature, is subsumed under the general law of the other cases" (GM V, 385).[14] The law of continuity thus permits seemingly paradoxical expressions: rest is motion, but the limiting case of ever-slower motions; a straight line is a circle, but the limiting case of circles of ever-greater radius; a point is a line, but the limiting case of ever smaller lines, etc. Writing to Varignon in 1702, Leibniz spoke of the law as "taking equality for a particular case of inequality, and rest for a

---

[13] One might note, in passing, that this is precisely the kind of argument that Newton attempts when he bases his method of fluxions on "ultimate ratios of evanescent increments." The connection between Leibniz and Newton on this issue is explored in Arthur (2008).

[14] This is but one of several statements of the law. The original appearance to which Leibniz refers is in the *Nouvelles de la république des lettres* in July of 1687 (GP III, 53–54). In a manuscript known as *Cum prodiisset...* written around 1701, Leibniz stated the law as holding "In any proposed continual transition ending in any terminus, it is permitted to formulate a general reasoning in which the last terminus may be contained" (Leibniz 1846, p. 40). See Schubring (2005, pp. 174–186) for an interpretation of the role of the law in Leibniz's mathematics .

particular case of motion, and parallelism for a case of convergence, etc. supposing not that the difference of magnitudes that vanish is already equal to nothing, but that it is in the act of evanescing, and the same of motion, which again is not absolutely nothing, but something on the point of being nothing. And if someone is not content with this, we can make him see, in the style of Archimedes, that the error is not at all assignable and cannot be given by any construction" (GM IV, 105).

The acceptability of fictional or ideal infinitesimals is therefore supposed to be guaranteed by the following sort of reasoning: having established some ratio or inequality holds for every finite increment of two quantities x and y, the law of continuity guarantees that this ratio or inequality holds even as the quantities evanesce. The derivative, thus understood, is a ratio between two evanescent increments of the ordinate and abscissa of a curve, whose relationship is given by the equation defining the curve and which determines the slope of the tangent throughout the curve. Likewise, the integral will be an "inverse tangent" construction, built up as the limiting case of approximations whose difference from the area enclosed by the curve evanesces. It is here that we can see the outlines of the "model-theoretic" strategy for vindicating the fictional infinitesimal. The introduction of the fiction is licensed by the law of continuity, and Leibniz held that this law would guarantee that any result obtained by using the fictional elements such as dx or dy would still be true in classical, finitistic mathematics. Leibniz's reference to Archimedean procedures thus seems intended as a means of justifying the law of continuity by claiming that any supposed error can be made to vanish.

Unfortunately for Leibniz, there is a good deal more work required to keep the law of continuity from delivering false results. As is clear from his argumentation in the justification of the calculus by ordinary algebra, Leibniz had no qualms about evaluating the expression $0/0$ as 1, and similar difficulties beset his understanding of the law of continuity. Consider, for instance, the following supposed application of the law of continuity: take $0 < x < 1$ and let x tend toward 0. As x gets smaller, the inequality $x^2 < x$ holds for all positive values of x on the way to zero; thus, applying the law of continuity, we seem entitled to conclude that $0^2 < 0$. These and similar difficulties can only be overcome by sharpening the formulation of the law of continuity in ways that specify when a "passage to the limiting case" is permissible. Doing that, however, would again require something very much like developing the apparatus of sequences, limits, and convergence conditions familiar from the later "rigorizations" of the calculus due to Cauchy and Riemann. Thus, his attempt to base the fictional infinitesimal on principles drawn from ordinary algebra left Leibniz still short of the desired goal.

4   Conclusion: What is the Leibnizian Standard of Rigor?

Let me hasten to a conclusion. I claim that Leibniz clearly committed himself to the notion that infinitesimals are fictions that are "ideal" from the mathematical point of view and need not be taken as genuine mathematical entities. In his embrace of the fictional infinitesimal, Leibniz cautioned that an excessive reliance on the traditional criteria for rigorous demonstration might impede mathematical progress, on the grounds that the results obtained by the calculus were sufficiently important that it would be irrational to demand that they be derived by the strict proof procedures familiar from classical geometry. Thus, although he gave lip service to the traditional notion that all of mathematics must rest on clearly evident first principles, Leibniz seems to have thought that actually working out these foundational principles was a matter that could be postponed to another, bleaker day. Moreover, when he undertook to justify the infinitesimal, either by showing that it could be eliminated in favor of exhaustion proofs, or by showing that recourse to such "ideal" elements would never lead us astray, Leibniz's efforts fell consistently short of the mark. In the *Arithmetical Quadrature* Leibniz's attempt to eliminate infinitesimal methods in favor of Archimedean exhaustions failed to achieve the desired generality, while in his "Justification of the Infinitesimal Calculus by that of Ordinary Algebra", his argument makes recourse to a continuity principle that needs some significant sharpening to avoid inconsistency.

Still, there can be little doubt that Leibniz had unshaken confidence that his calculus could be adequately justified, even if he could not carry out the justification himself. He never—so far as I know—suggested that the calculus offered only a method of approximation and not a means to demonstrate exact results. Nor did Leibniz embrace anything like a mathematical instrumentalism, regarding a mathematical theory as either false or meaningless, but acceptable because it reliably delivers correct results. In the end, Leibniz appears as an irrepressible mathematical optimist. Convinced that everything—but especially the realm of pure mathematics—was rationally structured, Leibniz declared to Varignon that everything worked out as if infinitesimals were "perfect realities", even if the truth of the matter is that there are no such things. A benevolent God, after all, would not strongly incline hard-working mathematicians to accept the calculus as true only to disappoint them with some nasty contradiction in their later results. In fact, I suppose that Leibniz would have embraced D'Alembert's famous dictum regarding the foundations of the calculus, which he is said to have repeated to his students some decades after Leibniz's death: "Allez en avant, et la foi vous viendra."

References

Arthur, Richard. 2008. Leery Bedfellows: Newton and Leibniz on the status of infinitesimals. In *Infinitesimal differences,* eds. Goldenbaum and Jesseph, Berlin: De Gruyter 7–30.

Barrow, Isaac. 1860. *The mathematical works.* 2 vols, ed. William Whewell. Cambridge: Cambridge University Press.

Bos, H. J. M. 1974. Differentials, higher-order differentials, and the derivative in the Leibnizian Calculus. *Archive for History of the Exact Sciences* 14:1–90.

Bos, H. J. M. 2001. *Redefining geometrical exactness: Descartes' transformation of the early modern concept of construction.* Berlin: Springer Verlag.

Dijksterhuis, E. J. 1987. *Archimedes. Transl. C. Dikshoorn, with a Bibliographic essay by Wilbur Knorr.* 2nd ed. Princeton: Princeton University Press.

Euclid. [1925] 1956. *The thirteen books of Euclid's "Elements" translated from the text of Heiberg*. Ed. and transl. Thomas L. Heath. 3 vols. Cambridge: Cambridge University Press, (New York: Reprint Dover).

Goldenbaum, Ursula, and Douglas Jesseph, eds. 2008. *Infinitesimal differences: Controversies between Leibniz and his contemporaries*. Berlin: De Gruyter.

Hankinson, R. J. 1995. Philosophy of science. In *The Cambridge companion to Aristotle*, ed. Barnes Johnathan, 109–139. Cambridge: Cambridge University Press.

Hobbes, Thomas. 1656. *Six lessons to the professors of the mathematiques*. London.

Hofmann, Joseph E. 1974. *Leibniz in Paris 1672–1676: His growth to mathematical maturity*. Cambridge: Cambridge University Press.

Ishiguro, Hidé. 1990. *Leibniz's philosophy of logic and language*. 2nd ed. Cambridge: Cambridge University Press.

Jesseph, Douglas M. 1998. Leibniz on the foundations of the calculus: The question of the reality of infinitesimal magnitudes. *Perspectives on Science* 6:6–40.

Knobloch, Eberhard. 1994. The infinite in Leibniz's mathematics: The historiographical method of comprehension in Context. In *Trends in the historiography of science, Boston studies in the philosophy of science*, eds. Kostas, Gavroglu, Christianidis Jean, and Nicolaidis Efthymios, vol. 151. Dordrecht: Kluwer.

Knobloch, Eberhard. 2002. Leibniz's rigorous foundation of infinitesimal geometry by means of Riemann Sums. *Synthese* 133:59–73.

Leibniz, G. W. 1768. *Opera Omnia*, ed. L. Dutens, 6 vols. Geneva.

Leibniz, G. W. 1846. *Historia et Origo calculi differentialis a G. G. Leibnitio conscripta*, ed. C. I. Gerhardt. Hanover.

Leibniz, G. W. 1993. In *De quadratura arithmetica circuli ellipseos et hyperbolae cujus corollarium est trigonometria sine tabulis*, Ed. with commentary by Eberhard Knobloch, Vandenhoeck and Ruprecht. Göttingen.

Leibniz, G. W. 1995. *La naissance du calcul differential: 26 articles des "Acta eruditorum"*. Ed. and transl. Marc Parmentier, Vrin.

Levey, Samuel. 2008. Archimedes, infinitesimals, and the law of continuity: On Leibniz's fictionalism. In *Infinitesimal differences*, eds. Goldenbaum and Jesseph, Berlin: De Gruyter 95–106.

L'Hôpital, G. F. A. 1696. *Analyse des infiniment petits pour l'intelligence des lignes courbes*. Paris.

Mancosu, Paolo. 1996. *Philosophy of mathematics and mathematical practice in the seventeenth century*. Oxford: Oxford University Press.

Parmentier, Marc. 1995. L'Optimisme mathématique. In *La naissance du calcul différentiel: 26 articles des "Acta eruditorum"*, ed. G. W. Leibniz Vrin, 11–52. Paris.

Pascal, Blaise. 1963. *Œuvres complètes*, ed. Louis Lafuma. Paris: Éditions du Seuil.

Roero, C. S. 2005. Gottfried Wilhelm Leibniz, First three papers on the Calculus (1684, 1686, 1693). In *Landmark writings in western mathematics: 1640–1940*, ed. I. Grattan-Guinness, 46–59. Amsterdam: Elsevier.

Rolle, Michel. 1703. Du nouveau systême de l'infini. *Histoire et Mémoires de l'Academie Royale des Sciences*, 1:312–336.

Schubring, Gert. 2005. *Conflicts between generalization, rigor and intuition: Number concepts underlying the development of analysis in 17th–19th century France and Germany*. New York: Springer.

Wallis, John. 1693–1699. *Johannis Wallis. S. T. D.... Opera Mathematica*. 3 vols. Oxford.

Whiteside, D. T. 1960–1962. Patterns of mathematical thought in the later seventeenth century. *Archive for History of the Exact Sciences* 1:179–338.

Index

A

Académie Royale des Sciences, 8, 11, 194
Action, 26, 81, 153
Activity, 85
Aiton, Eric, 81
Algebra, 4, 14, 15, 42, 66, 197, 203
Analysis
  logical, 50, 54, 61, 69, 71
  of notions, 52, 54, 55, 56, 57, 65, 66, 70
  of truths, 55
  situs, 5, 49
Apollonius, 196
Application
  of combinatorics, 26
  of mathematics, 24, 25, 34, 44, 202
Approximation, 32, 67, 117, 130, 200, 204
Aquinas, Thomas, 23
Archimedes, 31, 32, 33, 36, 43
Aristotle, 23, 28, 189, 190
Aristotle's wheel, 172
Arithmetic, 14, 17, 56, 77
Arnauld, Antoine, 7, 151
Arthur, Richard, 19, 142, 146, 185
Art of discovery, 43, 194, 195, 196
Atomism, 23, 26, 28
Axiom, 9, 12, 60, 196, 197
  Euclid's axiom (Part-Whole Axiom), 20

B

Bacon, Francis, 17, 76
Barrow, Isaac, 18, 111, 112, 125
Bassler, O. Bradley, 185
Bayle, Pierre, 26, 202
Beeley, Philip, 16, 185
Belaval, Yvon, 76
Benardete, José, 142
Bernoulli, Johann I, 46, 145, 194
Bertoloni-Meli, Domenico, 81

Bijection principle, 157, 163, 165, 172, 177, 180, 181
Bodenhausen, Rudolph Christian von, 46
Body, 13, 29, 79, 154, 180
Boineburg, Johann Christian von, 7, 24
Bos, Henk J.M., 18, 20
Bourguet, Louis, 82
Boyle, Robert, 125
Breger, Herbert, 144
Brown, Gregory, 142, 150
Brown, Richard C., 142
Brunschvicg, Léon, 6

C

Cantor, Georg, 137
Carcavi, Pierre de, 7
Cardano, Girolamo, 9, 23
Cassini, Giovanni, 7
Cassirer, Ernst, 5
Cauchy, Augustin Louis, 146, 200, 203
Cavalieri, Bonaventura, 13, 18
Ceulen, Ludolph van, 32
Chapelain, Jean, 32
Character
  alphabet of human thoughts, 15, 17, 51, 53
  universal, 14, 15
Characteristica universalis, 51, 65
Characteristic triangle, 78, 111, 117, 118, 124
Child, James-Mark, 117
Clarke, Samuel, 82
Class, 144, 159, 161
Colbert, Jean-Baptiste, 8
Collins, John, 126, 127
Combinatoria
  ars combinatoria, 15, 17
  combinations, 50, 61
Combinatorics, 26, 189
Computation, 50, 52, 61

© Springer Netherlands 2015
N. B. Goethe et al. (eds.), *G.W. Leibniz, Interrelations between Mathematics and Philosophy*, Archimedes 41, DOI 10.1007/978-94-017-9664-4